I0494702

JOURNEY INTO THE

GIANT SELENITE

CRYSTAL CAVES

OF MEXICO

(Edition II)

The Largest Crystals Discovered on Planet Earth

Leela Hutchison, G.G.

LEELA HUTCHISON

Copyright © 2016 Leela Hutchison
All rights reserved.
ISBN-13: 978-1537479736
ISBN-10:1537479733

To my parents who were determined my birth would happen in El Paso, Texas

NOTES TO THE READER

The adventures I experienced, remembered and retell in stories are shared with you, the reader, as a reminder of how the journey of our life expresses itself. By understanding the clues given by our experiences, we can foretell a chosen path or choose another. My intention here is to bring enjoyment and contemplation. May this book inspire you to create new adventures in your own life, at any age!

FORWARD

After several years of sharing the earliest images of the giant crystals of Mexico to audiences around the United States, Central and South America and Europe, time was overdue to write about my personal journey.

It was with a high degree of resistance that I disregarded the promptings of many to recount what I saw, felt and experienced when I entered those bubbles inside bedrock deep down in the earth.

I spent years attempting to pierce into a deeper realization of what these crystals purpose may be other than just a chance of perfect conditions by mother nature to create giants out of hydrated calcium sulfate and water.

Fifteen years later, my research into the deserts of the southwestern United States and Mexico has lead me to some interesting theories, ideas and conclusions. Those thoughts will be shared in another book forthcoming.

Learning how to write a book about something I was so completely passionate and curious about should have been easier and probably is to some. There were many events in my life that I believe were in preparation leading up to my invitation to explore the caves and I share them here. I believe they are all congruent with the larger subject at hand and that is to tell the story of what I witnessed inside those mysterious giants of crystal caves.

My intention of course, is to have you, the reader, join me and enter into those caves as though you were there yourself. I am here to retell an experience first hand that very few will ever have the privilege to share.

The mine has flooded the giants' crystals as of fall, 2015. I consider myself fortunate to have had the chance to explore the crystal caves as now they are forever out of reach and lost to the world. Those who own the silver mine have no intention of preserving their accidental discovery to share with future generations. They do not earn any money (from the crystals), and when the financial situation of the mine changed, as predicted, it closed its operations. The water pumps were removed and the caves are now flooded.

CHAPTER 1

The Feather River, Crystal Mountain and Plumas County, California
August 1998

I can't breathe this dust in any longer as I clenched my teeth. I closed my eyes for a moment while sucking the dirt and air in my lungs. I wanted to call out for help but held back and did my best to stay calm and in control. The falling rocks, dirt and dust from the small cramped hole underneath the massive oak tree were choking me. My face was less than an inch away from a dirt wall. I felt like I was going to panic any second from the feeling of intense claustrophobia. And what then? I desperately wanted to move forward quickly but feared I would be permanently stuck.

I was beginning to feel stricken with paralysis with one leg and arm just outside of the hole and the rest of me squeezed in between a rock face and a wall of dirt. Even though it was mid-day, it was damn dark in here. I needed my flashlight to see where this tight squeeze was going to lead me.

I had my coveralls on which were hot and itchy, like sand paper. Armed with a large screwdriver and a hammer in one hand, my fingers weren't doing me a lot of good feeling my way in the dark.

Now how was I going to use these simple tools and see where to dig in the first place?

Mike and Ray were like kids on a fun day trip excursion while I was sweating bullets wondering if I was ever going to see daylight again. They didn't pay the slightest bit of attention to my predicament and I didn't blame them.

What they saw had a firm grasp of their complete attention. In a chamber 15 feet below, there were big crystal points poking through the mud and dirt wall waiting to be pried out. There were too many to count. "Oh man, do you believe this?" Mike shouted. "Come on Leela, get on down here, this is incredible!" They estimated there were more than enough to fill both of their backpacks.

Fat chance I mumbled resenting them both for their fierce bravery and stupidity. What if a rock fell out from the wall and crushed them both?

But I was already envious of what they might find in the way of quartz crystals at the bottom of the tree roots not too far below me. I could only imagine that sight at this point and was starting to regret not moving fast enough and getting down there as quickly as they did.

Of course I was wondering if I would get my fair share before they had grabbed all of them. So I laid there on my back - stuck, and afraid to move my jammed body forward but also not wanting to retreat. I breathed slowly while listening to those guys squeal with joy.

One thing I knew for sure as I was thinking about them, they would have to crawl up the same way I was trying to get down there to meet them. At least I wasn't alone that meant I was not going to die in here. I thought, surely they would rescue me if I couldn't dislodge myself.

While I was waiting for them to come back up to find me, I was thinking how did my enthusiasm once again, lead me to this fun little adventure?

My friend Mike and I had taken a journey to the Grand Canyon a couple of years ago. We hiked down to Phantom Ranch and back up the south rim on the Bright Angel Trail in less than a day. We were proud of this accomplishment and promised we would have more adventures together.

One day, when we were back in Marin County, he had called me to say that his buddy and co-worker, Ray, a navy veteran, and expert in marksmanship, had just returned from Plumas County. He had been up at Crystal Mountain near the town of Quincy and came home with a backpack full of quartz crystals that he had dug out of the mountain.

Mike asked me if I wanted to see the crystals that Ray had given him? I jumped at the chance and we arranged a time to meet so he could bring over his collection. When I looked over the treasure trove of glittering, sparkling crystals, I was more than fascinated that these were freshly dug out of a mountaintop by his friend in the past few months.

I asked Mike to introduce me to Ray and to please ask him if he wouldn't mind taking another trip up there with the three of us before the end of summer. I wanted to see the source of these beautiful pieces of quartz crystal. I could take my own vehicle and camping gear and follow them there if he would consider it. Besides, I liked the idea of being as independent as possible so that I could leave when I wanted to.

Ray agreed to meet me and to take me up to Crystal Mountain with the both of them a few weeks later. He seemed a little quirky to me but highly intelligent. I found out later he used to live in the gold country in the Sierra Nevada.

Most people who live there are very independent, unique and eccentric. Some are hiding from the law and are fierce freedom lovers. I met one miner a couple of years earlier that carried a pistol in a holster on his hip and used a cable and a bucket to transport himself across the Yuba River where he had staked a gold claim.

This was gold country. One thing I discovered living in northern California and spending time in the Sierras was that where there is gold, you will find rock quartz and crystals. But where there are quartz crystals, you won't always find gold!

I realize now, Ray was never after the crystals. He and his buddies were searching for gold deposits in the mountains of the Sierra Nevada. That is probably why he consented to another crystal dig and allowed me to come along. He already knew there was no gold vein there to be found underneath that beautiful oak tree and no chance of exposing a secret that he would want to keep for himself.

The location for the town of Quincy was beautiful. There were several streams, creeks and rivers with clean air produced by the hundreds of thousands of pine trees in the area.

Quincy was a gold rush community established in 1858. There hadn't been a fire in the forest or mountains for over 50 years and there had been so much rain back in 1998 that the mountains were a lush green paradise with the Feather River running its sparkling course down into the Sacramento Valley.

As I choked on another breath of dust, I was thinking about how difficult it was to get here in the first place.

We had to take a very old logging road that hadn't been used in 50 years. The rains had gutted out the road so badly that huge tree trunks were sawed and cut down and drug into place by chains by either the Forest Service or a private logging company to replace the gaping ruts that could force a logging truck to roll down the mountain.

After a very bumpy four-wheel drive, we found an indicator where to stop our truck on the old path. Ray cut the engine. I looked around for a landmark so that I would remember how to come back here someday but there was nothing really to indicate where we were on the old logging mountain road.

It had already been a long drive and we hadn't even begun the dig yet. Ray pulled out coveralls and tossed them towards us. "Here, put these on" he commanded. We didn't argue or say a word about it and did as he demanded. Then the fun started. With empty backpacks and tools in tow, we had to climb up the steep mountainside using our hands and feet to get to the top.

Once we plateaued, out on the other side of the peak we could see all the way down to where the Feather River was coursing through the mountains. What a beautiful spot I said to no one as I marveled over Mother Nature. And then I saw her! A gorgeous Oak Tree that looked more like the Tree of Life with extraordinary laminated leaves, stems and branches that seemed to have living arms instead of tree limbs. She seemed to be welcoming us as the breeze rippled her leaves.

I gasped when Ray said, "That's it! That is where we go in!" I almost couldn't grasp what he was saying. Underneath her tree roots on the rocky slope, was a hole dug that went so deep that at the bottom there had been a chamber where you could actually stand up.

To take my mind off waiting for those guys, I fumbled with the flashlight to see well inside the hole. Praying it wouldn't roll away from me, I thought this would be a good way to kill some time if I could wave the light beam into the darkness and that is when I saw something.

Tiny diamond like features were flickering as though there were on and off switches for glittering lights. There were hundreds of points of light, directing their energy right into my eyes.

I felt a surge of excitement. I waved my flashlight again and saw them. Hanging from the top of the roof of the man made hole, there were tree roots, rocks, mud, water droplets and crystals! Crystals! I was looking at a giant jewelry box full of shiny objects and I was feeling giddy.

I was born and raised as a girl, and even though I acted more as a tomboy with the neighboring boys, you couldn't drag me away from anything that glittered. And that is the story I hear from just about every woman I know. We are crazy about everything that sparkles, glows, glitters and glistens.

I have had costumes, shoes, jewelry, rhinestones, head pieces, eye masks, purses, bags, pillows, curtains, drapes, twinkle lights and just about everything else that would catch your eye as a kid.

My eyes were transfixed. There were little ones and big ones and clusters, and perfect points and little castles, clear and milky and translucent with some crystals that were dripping drops of water. I saw them wedged in clusters in numerous pockets with little tree roots sticking out from the matrix.

This was far beyond my imagination about any rock I ever picked up off the ground in the deserts of the southwest or in my favorite childhood playground: McKelligon Canyon located in the Franklin Mountains of El Paso, Texas.

I had certainly bought quartz crystals before but I didn't have a clue as to how they were formed or how they were transported to some rock hound's table to sell to his customers. My mind had a million questions ready to spring on Ray or somebody who would be able to understand my complete amazement, inexperience and lack of knowledge. At this moment in my life, I didn't know how to pry one crystal away from its matrix of earth and mud.

Now I was eager beyond claustrophobia, exhaustion or heat. Damn the dust. I am going for it even if I only have one hand and arm to dig them out.

That screwdriver got a workout, digging and prying and using the claws of the hammer to drag them towards me. I dug out as many as I could and let them roll out and down the slope, hoping they wouldn't be lost or broken.

It was only when I could not reach one more sparkling fresh crystal that I gave up. Besides, how many of these beautiful little crystals did I need in my clutches?

So, I squeezed and squeezed and backed myself out of the hole and scrambled down the slope. I saw the last rays of light from the sun begin to fade and turn the river far below into a silver and orange ribbon.

I looked around and took a deep breath. I realized I was covered in dirt from my face down to the dirt filled in my boots.

I looked for my cache of virgin crystals that were strewn on the ground around me and I couldn't believe what I saw. They were coated in a very thin layer of dark black dirt. Now how could that be? They were crystal clear under the tree.

But wait a minute. I looked closer and what I saw amazed me. It wasn't flat black clay/mud but a kaleidoscope of beautiful metallic colors of pink, copper, amethyst, cobalt blue, and gold that coated the crystals.

I didn't know until years later, that is how the rainbows inside crystals get formed as well: from the water.

The crystals were damp from the water and minerals and silicon di-oxide dripping off the tips of the roots from the oak tree. They were living crystals still in a growth phase. And then right before my eyes, the black film dried into a light brown coating on the crystals.

Miners and rock hounds use Oxalic acid to clean the residue from crystals and bring them to gem and mineral shows around the world to sell as perfectly cleaned crystal and or white points.

I realized immediately something that was really important: these crystals had never seen the light of day in their entire growth period nor had another person touched them. Ever!

Ray and Mike were not far behind now as I heard them moving up through the hole and out.

Their treasures were undeniably beautiful. Large clusters with perfect points and some single points those were intact as well. They had both of their backpacks filled with them. I was amazed at the yield of crystals we were able to obtain.

I came away from that beautiful oak tree on that hot summer day with a backpack of my own crystals. And surprisingly they seemed to be talking to me!

These crystals were happy to come home with me on one condition. The message was that I was to give all of them away to people, especially to strangers, except for three that I was allowed to keep.

I drove home that night back to Marin County in a dreamlike state. How many people in this world get to have the opportunity that I just experienced? I was starting to realize one thing on this journey through life: Every step leads to the next. There is nothing random or by chance. And more importantly, there are guardians or angels along the path with us, assisting us with our decisions and giving us counsel. That weighed deeply with me as I contemplated this over and over again as I drove home.

That was 1998 and I was just beginning to ask the question: Where is this experience leading me? Why had I been underground in a dark sunless environment? What about Crystals? What is their importance?

I was really curious as to what would be my next step. I knew there was a purpose to this experience the three of us just had. I was wondering what it would be. As beautiful as the crystals were, I never dreamed I would be in a black hole again or in a place that would be considered the most hostile environment on the planet or that people would lose their lives while trying to steal crystals out of there.

Little did I know I would forget about this experience (nor link it to any past experiences) in less than four years of time. I was unaware this event had been preparing me on many levels by leading me to one of the greatest adventures in my life.

CHAPTER 2

White Sands National Park, New Mexico
1950's

I was born in El Paso Texas, the youngest of three children. My mother was from Pennsylvania and my dad was from the crystal fields of Arkansas. What I was going to learn about Arkansas in my future would help me understand my great interest and connection to geology, minerals and crystals.

The border town of El Paso was a rough and scary place to raise a family. There were over 350,000 people living there in 1970. On the other side of the United States border was the City of Juarez, in the country of Mexico. It would later be known as one of the most dangerous cities in Mexico for the drug cartel's illegal activity of smuggling drugs into the U.S.

It was here that I began in earnest my life long devotion to exploring Desert Mountains, caves and mysterious white sand. It was also here, where I learned to be more street-smart because of some very tough lessons about life. These lessons were to prepare me to be cautious and careful while taking risks in life.

As a very young child, I used to ask my parents "Where did I come from?" They would laugh and tease me and say, "Well, we found you under a rock." I would look completely dumbfounded at them, as my little mind couldn't comprehend what they were trying to tell me. This was serious business and I pestered them constantly about the origins of this rock.

As a four year old, I considered the truth about this rock identity.

But then, what kind of rock was it? Where was it? Did we have to drive to find it so I could see it? Or was it in our backyard? Had it been at the hospital? Just where would I find this rock?

This line of questioning brought the entire family in on it, so there was no use asking my older brother and sister if they knew anything more about the origins of this rock. They thought it was quite hilarious and they insisted as well that I was indeed found underneath a rock with no further details given to me.

I have had a good laugh about all of this over the years as I grew up and fell in love with rocks, crystals and gems. Maybe that is why I was in a constant search for beautiful, sparkly little rocks that I would find up in the canyon in the Franklin Mountains that graced the western horizon of my hometown of El Paso. In addition, the fossils I found in the deserts southeast of El Paso near Carlsbad Caverns, New Mexico would also be of great interest and a mystery to me.

I used to love going to McKelligon Canyon up in the Franklin Mountains in El Paso, for picnics with my family and exploring the steep rocky slopes. It seemed to me that if I kept searching, hiking higher up on the mountain, I was going to find that one special rock that I was born under.

However, I didn't find that rock anytime soon and over time, my fierce curiosity looking for the next canyon or mountaintop to explore continued to grow.

I remember my first visit to White Sands National Monument in southern New Mexico, located 30 miles from El Paso, Texas. It was here that I would realize the connection with selenite to the giant crystal caves in Naica.

The largest gypsum dune field in the world is located at White Sands National Monument in south-central New Mexico. This region of glistening white dunes is in the northern end of the Chihuahua Desert within an "internally drained valley" called the Tularosa Basin. The monument ranges in elevation from 3890' to 4116' above sea level. There are approximately 275 total square miles of dune fields here, with 115 square miles (about 40%) located within White Sands National Monument. The remainder of the dunes is on military land (White Sands Missile Range) that is not open to the public.

This dune field is very dynamic, with the most active dunes moving to the northeast at a rate of up to 30 feet per year, while the more stable areas of sand move very little. The pure gypsum (hydrous calcium sulfate) that forms these unusual dunes originates in the western portion of the monument from an ephemeral lake or playa with a very high mineral content. As the water evaporates (theoretically as much as 80" per year), the minerals are left behind to form gypsum deposits that eventually are wind-transported to form these white sand dunes. Many species of plants and animals have developed very specialized means of surviving in this area of cold winters, hot summers, with very little surface water and highly mineralized ground water.

It had been an early Saturday morning when the family headed out of town by automobile. I had just finished my first year of elementary school and it was the beginning of summer.

We drove a long stretch across a dry and thirsty desert-landscape that seemed to me a very lonely place as I looked out of the station wagon's windows. There were abandoned buildings along the old state highway 70, such as an old café, with broken windows and painted signs that had letters missing, bleached out from intense heat of countless summers of desert sun.

"What in the world could be more interesting than rocks and mountains I asked myself?" It was obvious that we weren't driving towards any mountains. I was sure that this outing was not going to be any fun at all as I was looking at all that flat desert passing by us.

At the entrance, the Park Ranger handed us a map and told us that we were not allowed to cross any boundaries over into the missile range's portion of the sands. That was strictly forbidden and with legal consequences of trespassing.

I had overheard my dad saying years later that was where the famous German Scientist, Werner Von Braun was brought to in the U.S. to help with furthering the space race back in the 1950s. He was an aerospace engineer and space architect credited with inventing the V-2 rocket for Nazi Germany and the Saturn V for the United States. He was one of the leading figures in the development of rocket technology in Nazi Germany, where he was a member of the Nazi Party and the SS. Following World War II, he was moved to the United States, along with about 1,500 other scientists, technicians, and engineers, as part of Operation Paperclip, where he developed the rockets that launched the United States' first space satellite Explorer 1, and the Apollo program manned lunar landings.

There were a lot of strange lights we would see from time to time when visiting Las Cruces, coming out of the Organ Mountains where the Range was next to. There were a lot strange booming noises as well. Probably from the underground bomb testing and ground to air missile fire testing as early as the 1940s.

The Park Ranger gave strict instructions. We were to travel only on the designated road into and out of the Park. Other than that, the giant mounds of sand were free to explore and play in. What was so forbidden here that we were not allowed to see I had wondered? Later, it would be fragmented parts of blown up missiles that had landed in the park.

I was a little confused however when everywhere that I looked I saw nothing but this pure white stuff. I asked my mom if it had snowed out here and if we were going to need our jackets? She laughed and said "Of course not. Its white sand not snow!" Well, I wasn't too sure about that. This place was huge and the sand went on for miles and miles. The roads were covered in this white sand just as snow would be.

We got out of the car with our picnic basket and placed it at one of the tables. It was warm outside, a huge contrast from the air-conditioned wagon we had travelled in. I was starting to get excited. This sand was absolutely brilliant and sparkling like snow. We didn't have hats or sunglasses but we sure could have used them.

I couldn't see over the tall mounds and that meant I had some climbing to do. What would I find over the next sand dune? We spent the entire day playing in that sand. I rolled from the top of the mounds down to the bottom, getting sand in my hair, eyes, nose and mouth.

It would be many years later after 2000, when the giant crystals had been discovered, that I had started my research in and around the area of the Chihuahua Desert and started connecting the dots. The massive white sand dunes resulted from gypsum selenite crystals being worn down by wind and water erosion over hundreds of thousands of years.

Gypsum is rarely found in the form of sand because it is water-soluble. Normally, rain would dissolve the gypsum and carry it to the sea. The Tularosa Basin is enclosed meaning that it has no outlet to the sea. The torrential rains dissolved the minerals and gypsum from the surrounding San Andres and Sacramento Mountains and were trapped within the basin. Thus water either sinks into the ground or forms shallow pools that subsequently dry out and leave gypsum in a crystalline form, called selenite, on the surface.

Groundwater that does flow out of the Tularosa Basin flows south into the Hueco Basin. During the last ice age, a lake known as Lake Otero covered much of the basin. When it dried out, it left a large flat area of selenite crystals that is now the Alkali Flat. Another lake, Lake Lucero, at the southwest corner of the park, is a dry lakebed, at one of the lowest points of the basin, which occasionally fills with water.

The ground in the Alkali Flat and along Lake Lucero's shore is covered with selenite crystals that reach lengths of up to three feet (1 m). Weathering and erosion eventually breaks the crystals into sand-size grains that are carried away by the prevailing winds from the southwest, forming white dunes. The dunes constantly change shape and slowly move downwind. Since gypsum is water-soluble, the sand that composes the dunes may dissolve and cement together after rain, forming a layer of sand that is more solid and could affect wind resistance of dunes. This resistance does not prevent dunes from quickly covering the plants in their path. Some species of plants, however, can grow fast enough to avoid being buried by the dunes.

Unlike dunes made of quartz-based sand crystals, the gypsum does not readily convert the sun's energy into heat and thus can be walked upon safely with bare feet, even in the hottest summer months. In areas accessible by car, children frequently use the dunes for downhill sledding.

Because the park lies completely within the White Sands Missile Range, both the park and U.S. Route 70 between Las Cruces, New Mexico and Alamogordo are subject to closure for safety reasons when tests are conducted on the missile range. On average, tests occur about twice a week, for duration of one to two hours.

Located on the northernmost boundaries of White Sands Missile Range, is the Trinity Site. This is where the first atom bomb in the history of humankind was detonated in July 1945. A terrifying landmark in history, this place is where a handful of gleeful scientists ability to obliterate thousands if not millions of people with a death bomb of horrible proportions is looked upon with pride of the future of America's mighty military to stop Hitler.

What did this place look like a million years ago? Were there a multitude of small or big crystals living in the salt beds? What happened to all of the massive water containment? And what was under the dry inlet lake now?

All these questions were leading me to understand that there was a vast network of invisible gridlines that connected energy from one place to another. Also there is a massive aquifer of water underneath this endless desert. White Sands National Park indeed was connected to Naica where the giant crystals of selenite were discovered.

At the time of my exploration into the giant caves in 2001, I didn't know what gypsum was comprised of or that the selenite crystals are the crystallized version of gypsum – also known as Hydrated Calcium Sulphate.

There was a lot of information to process and it would take me years to learn much more about crystals and their habits.

Here was a perfect example of one step leading to the next. I was being bathed in selenite crystals at a very early age and White Sands National Monument was in my backyard - almost. And Naica, Mexico was only 200 miles south of El Paso.

I had the eerie feeling that my future self was intermingling with my past experiences leading me down a path to unravel the mysteries connected with selenite crystals.

CHAPTER 3

The Grand Canyon of Arizona
Spring 1986

After a few years of college, I moved to Arizona. Here was a place like none other with majestic canyons, lakes with strange looking Saguaro cactus and the incredible beauty of Desert Mountains and mesas.

I was blessed with a deep love for nature and a curiosity for adventure. This included the next mountain range or an Indian petroglyph hiding behind boulders. Climbing rocks became second nature for me.

It was a cool and crisp fall day in 1986 when Tomi, a girlfriend, convinced me to take a drive to the Grand Canyon and spend the night on the south rim.

I had been depressed for several months from dealing with the news of a terrible car accident that left that man I had fallen in love with paralyzed for life. He was never going to recover from this traumatic and devastating injury. The heartbreak I felt from knowing we would not be together was beyond my ability to cope with the situation.

Tomi had been encouraging me for weeks to get out and go explore nature and the outdoors of Arizona with her. She had planned a trip to the Grand Canyon on business and invited me to share her hotel room with her.

We had arrived there early in the afternoon and as we walked towards the canyon's rim and looked over and down the six thousand feet of desert terrain, we saw this very tiny trail that extended towards the river and dropped over a plateau into thin air.

I had asked her, "What in the world was that little trail leading towards, deep down in the canyon?" Tomi laughed and replied, "That is the Bright Angel trail that leads down to Phantom Ranch and the mighty Colorado River" grinning with a huge smile!

Having an intense curiosity about hiking that trail, I looked into finding out more of what this extraordinary place was about: The Grand Canyon is a big fissure in the Colorado Plateau. Geologically, it is significant because of the thick sequence of ancient rocks that is beautifully preserved and exposed in the walls of the canyon. These rock layers record much of the early geologic history of the North American continent.

Uplift associated with mountain formation later moved these sediments thousands of feet upward and created the Colorado Plateau. The higher elevation has also resulted in greater precipitation in the Colorado River drainage area, but not enough to change the Grand Canyon area from being semi-arid. The uplift of the Colorado Plateau is uneven, and the Kaibab Plateau that Grand Canyon bisects is over a thousand feet higher at the North Rim (about 1,000 ft. or 300 m) than at the South Rim.

Almost all runoff from the North Rim (which also gets more rain and snow) flows toward the Grand Canyon, while much of the runoff on the plateau behind the South Rim flows away from the canyon (following the general tilt). The result is deeper and longer tributary washes and canyons on the north side and shorter and steeper side canyons on the south side.

Temperatures on the North Rim are generally lower than those on the South Rim because of the greater elevation (averaging 8,000 ft./2,438 m above sea level). Heavy rains are common on both rims during the summer months. The Grand Canyon is part of the Colorado River basin that has developed over the past 70 million years.

Recent research states: Mystery of Grand Canyon's Formation Revealed! Live Science - April 28, 2011. The birth of the Grand Canyon and the Colorado Plateau through which it carved has been a geological mystery. Now a giant anomalous structure discovered on the underside of the plateau could shed light on how it was formed. Over the past 70 million years, and possibly quite recently, the relatively flat Colorado Plateau of the southwestern United States - a 130,000-square-mile (336,000 square kilometers) region that straddles Colorado, Utah, Arizona and New Mexico - rose up about 1.2 miles (2 km), was invaded by magma and was eroded away into deep valleys, creating a dramatic landscape including the Grand Canyon. This kind of behavior is more expected with mountain belts, not plateaus, and so these events have perplexed geologists for more than a century.

The canyon is the result of erosion that creates one of the most complete geologic columns on the planet.

The major geologic exposures in the Grand Canyon range in age from the 2-billion-year-old Vishnu Schist at the bottom of the Inner Gorge to the 230 million year-old Kaibab Limestone on the Rim. There is a gap of about a billion years between the 500 million-year-old stratum and the level below it, which dates to about 1.5 billion years ago. This large unconformity indicates a period of erosion between two periods of deposition.

Many of the formations were deposited in warm shallow seas, near-shore environments (such as beaches), and swamps as the seashore repeatedly advanced and retreated over the edge of a proto-North America. Major exceptions include the Permian Coconino Sandstone, which contains abundant geological evidence of Aeolian sand dune deposition. Several parts of the Supai Group were deposited in non–marine environments.

The great depth of the Grand Canyon and especially the height of its strata (most of which formed below sea level) can be attributed to 5–10 thousand feet (1,500 to 3,000 m) of uplift of the Colorado Plateau, starting about 70 million years ago. This uplift has steepened the stream gradient of the Colorado River and its tributaries, which in turn has increased their speed and thus their ability to cut through rock

Weather conditions during the ice ages also increased the amount of water in the Colorado River drainage system. The ancestral Colorado River responded by cutting its channel faster and deeper.

The base level and course of the Colorado River (or its ancestral equivalent) changed 5.3 million years ago when the Gulf of California opened and lowered the river's base level (its lowest point). This increased the rate of erosion and cut nearly all of the Grand Canyon's current depth by 1.2 million years ago. The terraced walls of the canyon were created by differential erosion.

Between 100,000 and 3 million years ago, volcanic activity deposited ash and lava over the area, which at times completely obstructed the river. These volcanic rocks are the youngest in the canyon.

Reading and researching about the Canyon was all I needed to peak my curiosity for the hike to the bottom where Phantom Ranch was and camp overnight next to the Colorado River. I immediately began to plan my first major hiking trip down the Grand Canyon with Tomi.

These hiking adventures were preparing me for a level of endurance I was going to need in the future and revealed to me a surge of determination I didn't know I was capable of.

That first trip down the Grand Canyon was a thirty-six hour event, where we hiked down and back out again after spending the night down at Phantom Ranch. That was 20 miles round trip and it took every ounce of energy out of me when we were crawling back up to the south rim from the trail.

We had started out at 6:00 am with only a backpack and wearing warm clothes at 7,000 feet. It was chilly. We had to carry loads of water and food with us that made our packs heavy. We were dreading the heat later on in the day and having to carry the load of extra heavy clothing but it was going to be essential to have when we returned at the end of our journey.

The going wasn't terribly difficult but my feet weren't accustomed to wearing shoes for that long of a period of time and the heat really made my feet sweat. There were the beginnings of blisters welling up on my heels.

The heat was increasing with every mile we hiked deeper into the canyon. We were both fantasizing about getting into the cold water of the Colorado River when we had made it down to Phantom Ranch.

Native American Indians used the site of the ranch. Pit houses and a ceremonial kiva dating from about 1050 have been found there. The earliest recorded visit by Europeans took place in 1869, when John Wesley Powell and his company camped at its beach. Prospectors began using the area in the 1890s, using mules to haul their ore.

It was nearing nighttime when we made it back up to the top of the Bright Angel Trail. There was a full moon and it floated above us like a giant hot air balloon right above our heads. It seemed as though we could reach up and touch the huge orb as it lit up everything in the canyon. The moonlight made the light shining onto the north rim appear as if it was just out of arm's reach instead of the 10-mile distance from rim to rim, as we looked northward from a south rim perspective. The visual indeed was very surreal after our mind-altering trek.

With sore muscles and a feeling of deep satisfaction, we drove home in silence to a night sky full of stars as we headed south towards Phoenix.

That would be the beginning of several journeys in the years to come that I would take into the canyon. I was more than curious about learning of its history and the strange occurrences that were reported there.

Such as the report by G. E. Kinkaid published in the Arizona Gazette on April 5, 1909: He was an explorer and hunter all his life, working thirty years for the Smithsonian Institute. Below are excerpts from his journal of his alleged adventures into the Grand Canyon.

"I was journeying down the Colorado river in a boat, alone, looking for minerals. Some forty-two miles up the river from the El Tovar Crystal Canyon, I noticed, on the east wall, stains in the sedimentary formation about 2,000 feet above the river bed. There was no trail to this point, but I finally reached it with great difficulty.

This cliff face is purported to be the location of the cave entrance to the mysterious underground citadel.

The entrance is 1,486 feet down the sheer canyon wall. Above a shelf, that hid it from view from the river, was the mouth of the cave. There are steps leading from this entrance some thirty yards to what was at the time the level of the river.

When I saw the chisel marks on the wall inside the entrance, I became interested. Securing my gun, I went in.

I gathered a number of relics, which I carried down the Colorado to Yuma, from whence I shipped them to Washington with details of the discovery. Following this, other explorations were undertaken. So interested have the scientists become, that preparations are being made to equip our camp for extensive studies, the number of archaeologists increasing to from 30 to 40.

From the long main passage, another mammoth chamber has

been discovered from which radiates scores of passageways, like the spokes of a wheel.

Several hundred rooms have been discovered, reached by passageways running from the main passage, one of them having been explored for 854 feet and another 634 feet. The recent finds include articles that have never been known as native to this country, and doubtless they had their origin in the orient. War weapons, copper instruments, sharp-edged and hard as steel, indicate the high state of civilization reached by these people.

The main passageway is about 12 feet wide, narrowing to nine feet toward the farther end. About 57 feet from the entrance, the first side-passages branch off to the right and left, along which, on both sides, are a number of rooms about the size of ordinary living rooms of today, though some are 30 by 40 feet square. These are entered by oval-shaped doors and are ventilated by round air spaces through the walls into the passages. The walls are about three feet six inches in thickness.

The passages are chiseled or hewn as straight as could be laid out by an engineer. The ceilings of many of the rooms converge to a center. The side-passages near the entrance run at a sharp angle from the main hall, but toward the rear they gradually reach a right angle in direction.

Over a hundred feet from the entrance is the cross-hall, several hundred feet long, in which are found the idol, or image, of the people's god, sitting cross-legged, with a lotus flower or lily in each hand. The cast of the face is oriental. The idol almost resembles Buddha, though the scientists are not certain as to what religious worship it represents. Taking into consideration everything found thus far, it is possible that this worship most resembles the ancient people of Tibet.

Surrounding this idol are smaller images, some very beautiful in form, others crooked-necked and distorted shapes, symbolical, probably, of good and evil. There are two large cactus with protruding arms, one on each side of the dais on which the god squats. All this is carved out of hard rock resembling marble.

In the opposite corner of this cross-hall were found tools of all descriptions, made of copper. These people undoubtedly knew the lost art of hardening this metal, which has been sought by chemicals for centuries without result.

On a bench running around the workroom was some charcoal and other material probably used in the process. There is also slag and stuff similar to matte, showing that these ancients smelted ores, but so far no trace of where or how this was done has been discovered, nor the origin of the ore.

Among the other findings are vases or urns and cups of copper and gold, very artistic in design. The pottery work includes enameled ware and glazed vessels.

Another passageway leads to granaries such as are found in the oriental temples. They contain seeds of various kinds. One very large storehouse has not yet been entered, as it is twelve feet high and can be reached only from above.

Two copper hooks extend on the edge, which indicates that some sort of ladder was attached. These granaries are rounded, as the materials of which they are constructed, I think, is very hard cement. A gray metal is also found in this cavern, which puzzles the scientists, for its identity has not been established. It resembles platinum. Strewn promiscuously over the floor everywhere is what people call "cats eyes', a yellow stone of no great value. Each one is engraved with the head of the Malay type.

Carved on all the urns, over doorways, and tablets of stone, are mysterious hieroglyphics, the key to which the Smithsonian Institute hopes to discover. The engravings on the tablets probably have something to do with the religion of the people. Similar hieroglyphics have been found in southern Arizona.

Among the pictorial writings, only two animals are found - one of them looking prehistoric.

The tomb or crypt in which the mummies were found is one of the largest of the chambers, the walls slanting back at an angle of about 35 degrees. On these are tiers of mummies, each one occupying a separate hewn shelf. At the head of each is a small bench, on which is found copper cups and pieces of broken swords. Some of the mummies are covered with clay and all are wrapped in a bark fabric.

The urns or cups on the lower tiers are crude, while as the higher shelves are reached, the urns are finer in design, showing a later stage of civilization. It is worthy of note that all the mummies examined so far have proved to be male, no children or females being buried here. This leads to the belief that this exterior section was the warriors' barracks.

Among the discoveries no bones of animals have been found, no skins, no clothing, and no bedding. Many of the rooms are bare but for water vessels.

One room, about 40 by 700 feet, was probably the main dining hall, for cooking utensils are found here. What these people lived on is a problem, though it is presumed that they came south in the winter and farmed in the valleys, going back north in the summer.

Upwards of 50,000 people could have lived in the caverns comfortably. One theory is that the present Indian tribes found in Arizona are descendants of the serfs or slaves of the people who inhabited the cave.

Undoubtedly a good many thousands of years before the Christian era, a people lived here which reached a high stage of civilization. The chronology of human history is full of gaps.

One thing I have not spoken of may be of interest. There is one chamber of the passageway that is not ventilated, and when we approached it a deadly, snaky smell struck us. Our light would not penetrate the room, and until stronger ones are available we will not know what the chamber contains. Some say snakes but others think it may contain a deadly gas or chemicals used by the ancients. No sounds are heard, but it smells snaky just the same.

The whole underground installation gives one of shaky nerves the creeps. The gloomy feeling is like a weight on one's shoulders, and our flashlights and candles only make the darkness blacker. Imagination can revel in conjectures and ungodly daydreams back through the ages that have elapsed till the mind reels dizzily in space" by G.E.Kinkaid.

What other secrets and strange phenomenon does the canyon hold?

CHAPTER 4

Sweat lodges and visions – Carefree, Arizona
December 1993

I experienced my first sweat lodge with fasting in the high desert near Carefree, Arizona in December of 1993. I knew this all day fast was going to be a stretch for me and out of my comfort zone.

Some people show up ahead of time for a sweat lodge, pay the fee and just be present when the ceremony begins, but my friends and I were going create this lodge by sticks and stones that we were to find in the desert. That was part of the process of purification as we prepared ourselves for the sweat.

We were up early on a Saturday morning and drove east towards Fountain Hills to find lava rocks in the desert hills that we could use for our fire pit.

We had found an area near the off-limits signpost of the sacred Red Mountain that was part of the Yavapai Indian Reservation. This place had enough cinder cone debris scattered everywhere we looked. We could quickly pick up the 40 stones and placed them in the back of the truck for heating the lodge as well as the fire pit.

Walking and hiking and picking up somewhat heavy stones on an empty stomach had me already feeling weak but we were just at the beginning of our adventure. At least there were four of us to do the gathering that made it easier.

We started driving with a truck full of lava rocks as we made our way eastward and then northward on a dirt road. We soon found an arroyo (a dry creek bed) with soft sand that we could use as a level location to place our soon to be built lodge upon.

After arriving, we took some old machetes out of the truck and started to walk out into the desert landscape and cut down ocotillo spines that were studded with very sharp needle-like thorns.

Wearing thick gloves and gently carrying the 6-foot spines, we used the knives to shave off the thorns and use the green and flexible spines as lodge poles that would support building our small dome-like structure. Here we would use blankets and black plastic to wrap the dome that would hold the heat and steam in our small lodge.

This wasn't an easy task for me because of the fasting and I found myself feeling a little light-headed in the sun and dreaming of being a Native American warrior. Carrying a machete in my right hand seemed as natural as anything to me. I had never held one in my hands before. I was wondering about past lives and if indeed I might have lived my life as a native warrior or lived in this area at another time.

Once we had built our dome, we started a huge fire of gathered wood that we placed on top of the lava rocks and let the fire intensify the heat in the stones.

We built a deep earthen pit inside our lodge to carry the rocks in with sacred deer antlers. We then placed them one by one so that the entire pit was inlayed with these super hot stones.

As we prepared to enter the lodge the sweat leader smudged each one of us with the smoke of burning sage, wafting the smoke over us with an eagle feather. We then crawled into the lodge in a sun-wise (clockwise) direction, bowing in humility to Great Spirit/Creator. Feeling in close contact with Mother Earth, we would take our place in the circle, sitting cross-legged and upright against the wall of the lodge.

We then started what we called the rounds. Silently, we asked for guidance for what it was that each one of us needed to see in our own personal lives. We began to peel the layers off like an onion of our psyche. It was deeply personal. No one took this as an easy process to express. It had been a physically taxing day for us to create the lodge. Each one of us genuinely wanted a sacred moment with the Divine or Great Spirit.

As we sat in silence, dripping sweat, a wooden bucket with ladle was filled with water and poured over the stones to create steam to further our purification. I started to drift deeper into my own awareness.

It had been a tough year for me. My identity as a commercial real estate broker had completely unraveled since my boyfriend had been paralyzed in his car accident in 1986. As my income dwindled, I had attempted and yet failed to be enthusiastic or successful at other jobs and employment that I had taken on. I was at my wits end trying to figure out any solution. I was depressed and had no desire to continue to make money my top goal in life any longer. Yet, I was still a long way from finding any answers to what would take its place or discovering my soul's purpose.

I had lived most of my life up to this point with little contemplation for spiritual pursuits. I longed for nature and she was my dearest companion when I needed nurturing. Yet, I continued attempting to figure everything out with my mind and it was getting me nowhere. It was at this point in time that I was close to giving up by not being able to solve my so-called problems around my identity crisis. If a sweat lodge could offer me the opportunity for insights or visions, I was all for it.

Thirty minutes in the lodge with blazing stones was an intense experience. When we could no longer tolerate the heat, one by one, we would crawl out into the cool winter air of the now descending sunlight and allow the sweat to evaporate. Then the next round would begin. Each one of us was determined to stay longer than the last round inside the lodge. Yet we found we could stay only a shorter time on the next round.

By the third round, I was starting to have a clear vision coming through. I will never forget what I saw. It was shown to me over and over again as I sat in meditation and/or silence with my eyes closed while I drew in slow even breaths against the agonizing heat.

I envisioned the fetus of a human baby emerge with its eyes closed, being lifted from the placenta and blood of a mother's womb. I realized this baby was me and that I was giving birth to a whole new version of me. My past was dying and my old identity was being shed like a snake's skin. It was a powerful vision and I was awestruck. I had never remembered having a vision before. I wondered where this vision was leading me.

None of us spoke of our personal journeys that day as we broke our fast after the fourth and final round. My friend just smiled at me, watching me break through some barriers that I had not even known existed inside of me. We then broke down our lodge making sure to leave no trace of our activity in the dry arroyo and headed back home to Phoenix.

Things started to move very quickly for me after that. In January, I visited good friends in southern California to journey on the inner planes with medicine to explore consciousness.

This showed me the level of neglect and self-hatred that I had no understanding of and the first glimpse of returning to an awareness of love and forgiveness for myself and for others.

I remembered driving back to Phoenix, crying non-stop all the way home. The tears continued for another week after I returned. I was deeply sad and upset. I was unable to find the truest answers as to why I was reacting this way. I knew at this point that I was not going to live in Phoenix much longer or stay on my old and familiar path out of habit. This identity of mine didn't feel authentic.

Within 2 months, I had put my things in storage and moved out of Phoenix. I traveled to northern California and made my first home in Sausalito. I arrived there on March 21, 1994.

This change in residing in a place other than the southwest would begin my first steps into meditation, studying esoteric teachings and journeying on the path of what I call the Mystic - one who seeks to find out real knowledge and wisdom. Something out of the ordinary was coming.

CHAPTER 5

Arrival in Sausalito, California
March 1994

It was a foggy and overcast day when I drove into the North Bay area of San Francisco. I had my car packed to the roof with clothes and boxes going to my cousin's terraced home in Hurricane Gulch, a neighborhood in the hills of Sausalito.

I had driven most of two days from Phoenix, Arizona to enter the land of misty fog, creating a blanket of silence over the city. The drive was mostly desert until I reached the East bay and headed westward towards San Francisco.

Seeing and feeling the moist air and fog rolling in from the Pacific Ocean, I rolled my windows up from the chill starting to creep into my bones. It would take me years to get use to the climate. It was cold and damp and the smell of soil, pine trees and mold filled the air as I parked my car at the foot of the hill. I climbed the steps and 68 wood stairs to arrive at the front door of my cousin's home. I counted those stairs many times as I made my way up and down the hills of Sausalito. It was a gorgeous view of the bay and surrounding hills with sailboats and yachts dotting the waterscape.

I wasn't used to the dampness and within weeks of moving in, my first real challenge presented itself with major health issues of the lungs. Not realizing at first, I was beginning to see that I had been covering up and masking certain health symptoms by living in an arid climate. But there was no escaping any of that now in this very damp and cool climate.

I had been sick off and on for months after arriving. With no new friends yet or a job, I dove into psychic studies and meditation, wrapped in blankets on the pullout bed in the tiny bedroom called an office. And when I felt good, I would explore and discover every hiking trail that was in the Marin Headlands and Open Space and the gorgeous mountain called Mt. Tamalpais. She would later become my outdoor home for 18 years and I would become one of her devoted guardians.

I realized later, that I was beginning to peel off layers of the onion, one by one - in my desire to wake up and be a conscious spirit living a human existence.

I had done so much to mask my deep emotional hurts and pain over the years. Yet, I was completely unconscious about any of this masking. I had a lot to learn and it would take years for me to start to balance out and stabilize on every level.

But first came so many issues of deep grief showing up for me in the way of bronchial illnesses such as colds and flu viruses and bronchitis. It was to the breaking point that I finally had a major breakthrough of understanding and release.

Once I had come out on the other side of understanding and clearing one major challenge, another one would emerge. These issues were stacked so closely together, it would seem that there was no time to recover and relax before another wave of emotional issues would begin.

On September 26, 1994, my father had died at age 71 of complications resulting from congestive heart failure. I remember walking down one of the streets in the financial district of San Francisco heading towards an interview for employment in the Embarcadero area when I received word from family.

This was devastating news for me as there was a feeling of an incompletion and closure that had not happened between us. I had to draw to me a level of composure and self-confidence that I clearly didn't possess. I sat in the lobby of an office building waiting for a senior vice-president to walk thru to greet me and bring me into an interview. Needless to say, I did not get that job but my father was watching out for me and in a matter of days, I received an offer to manage a portfolio of real estate holdings for a national bank, with their world headquarters located in San Francisco.

In one year, I had already gone through several clearings and changes. I still had not stepped into a more meaningful purpose through right livelihood/career and was dealing with the shadow side of my psyche. I was in need of crashing my corporate-like identity while acting in the world of commercial real estate.

I was given a substantial financial gift when things began to wind down in my job. The company let go of several outside contractors and went "in house" for employees to take over my position. My career as a commercial real estate manager/broker was about to end forever.

A phone call and an invitation to travel came to me in November of 1995, and I was soon to have my first introduction into the world of the mysterious Tarahumara Indians of Copper Canyon in the Sierra Madre Mountains of Mexico.

CHAPTER 6

The Barrancas de las Cobres and the Tarahumara Indians
November 1995

Unexplained mysteries of synchrony and their long-range effects can lead us to our purpose and/or our destinies. I was soon to find out how easily it was to underestimate those subtleties.

We were visitors on the edge of a massive canyon that seemed like an entire new earth plane in the Sierra Madre Occidental of Mexico.

Overlooking a flat and worn desert-like playing field, I was sitting on an old tree trunk with my friends, Gil and Esther from Arizona. We were squinting while wearing sunglasses, as we directed our attention southeastward into the morning sun that seem to laser a beam of searing heat down on us and the players below running around on that field. It resembled more of a pasture pen than what would be fitting for some of the greatest ultra long distance runners in the world.

The field had old wooden posts and rustic barbed wire containing it and this prevented the distracted ball players from falling over the edges into the steep canyon of Las Barrancas del Cobre or Copper Canyon below. Even though it was November of 1995, the sun was intense and brilliant on the high desert plateau with gusty winds flying up from the depths of the canyon.

There were 10 men in colorful plain shirts with shorts that resembled pinup diapers more than a version of athletic clothing. Peter Essick, a photographer for National Geographic and my two traveling companions sat next to me staring as if we were in a hypnotic state, watching the muscular and stalwart ball players kicking around a small hand made wooden ball with their bare feet. Their legs resembled the sturdiness of tree trunks-thick, dark brown legs, with scars and scratches that escaped none of the players. Their sandaled feet were truly dirty, with either a blackened big toe nail or the vestiges of something similar to a toenail that was mostly ripped off. This was a testimony of kicking a wooden ball around that felt as heavy as iron and as such was named ironwood.

Their athletic shoes for this dangerous country full of steep trails, snakes, cactus and sharp-edged rocks were nothing short of miraculous with no thought or fear of protection for any of these things.

Incredibly, their shoes were not shoes at all but sandals made of a thin layer of leather with a thin layer of shaved automobile tire glued to the bottom. Holding this flimsy support to their feet was a leather string between their toes and wrapped around their thick steel-like ankles.

What astonished us the most, was the surreal way these men seemed to float in the air for seconds on end without a of trace of gravity affecting them as they ran and jumped around the field kicking that ironwood ball around. What strength these people had.

We were guests and visitors here in the Indian village of Creel, Mexico, which sat on the very edge of one of the deepest canyons in the world. We stayed the night in an old adobe home that had a black wood stove in the middle of the one room. On top, was a black kettle of pinto beans that had been cooking all day. With a few flies buzzing around and laying their eggs, I was sure to be blaming that particular delicious dinner for a very upset stomach the next day.

This place was home for the Tarahumaras, or should I say the mysterious Tarahumaras who lived here. They were deep into a game that was similar to soccer, kicking an orange-sized ironwood ball instead of a larger soccer ball.

The Tarahumara, or Rarámuris, as they refer to themselves, are one of the largest indigenous groups currently in the Americas. The majority lives closer to towns but some live among the precipitous peaks of the Sierra Madre Occidental in the state of Chihuahua, Mexico. It is in the mountains, also known as the Sierra Tarahumara, that the traditional ways of language, dress, ceremonies, food gathering persist. In the cities, the Tarahumara have had to adapt to the way of the chabochi, or anyone who's not Tarahumara.

The Rarámuris are accomplished distance runners. Because of the rugged, remote, and high terrain they inhabit, they travel by foot much of the time and can trek long distances. They play a kick-ball game called rarajípari, during which they may run 12 to 120 miles, depending on the ability of the participants. The rules vary, but the gist of the game is that players run a set of laps kicking an ironwood ball. Whoever reaches the end point with the ball first gets the winnings, usually goods put in a pile before the race begins

I was fascinated with this small coarsely hand made ball, which comes from the desert tree: Olney, also known as desert ironwood. It is a very hard wood and heavy. Its density is greater than water and thus sinks; it does not float downstream in washes, and must be moved by current motion. Due to its considerable hardness, processing desert ironwood is difficult.

These are the greatest ultra long distance runners on the planet for endurance running. Able to run 100 miles or more at a given time, with no stopping for thirst, no resting after feeling winded with the exception of smoking an occasional cigarette. Nothing but relentless running, ascending and descending the steep, rocky and narrow trails of one of the deepest canyons in the North America, Copper Canyon. These Indians were made for this terrain!

This is the Barrancas de las Cobres of Mexico. It is a system of canyons consisting of six distinct canyons in the Sierra Madre Occidental in the southwestern part of the state of Chihuahua in Mexico. The overall canyon system is larger and portions are deeper than the Grand Canyon in Arizona. These inter-connected canyons are formed by six rivers that drain the western side of the Sierra Tarahumara Mountains (a part of the Sierra Madre Occidental). All six rivers merge into the Rio Fuerte and empty into the Sea of Cortez. The walls of the canyon are a copper/green color, which is where the name originates.

The Spanish arrived in the Copper Canyon area in the 17th century and encountered the indigenous locals throughout Chihuahua. For the Spanish, Mexico was a new land to explore for gold and silver and also to spread Christianity. The Spanish named the people they encountered "Tarahumara", derived from the word Rarámuris, which is what the indigenous people call themselves.

Some scholars theorize that the word may mean "The running people".

During the 17th century, the Spaniards in the land of the Tarahumara tribe discovered silver. Some were enslaved for mining efforts. There were small uprisings by the Tarahumara, but to little avail. They were eventually forced off of the more desirable lands and up into the canyon cliffs.

My eyes drifted looking over the vast canyon. I was looking at the many trails that were leading off and down the steep and narrow walls, wondering where they lead. Our time here was short, just a couple of days as we explored the river at the bottom of the Canyon for a night and then stay with a family in Creel, a town on the edge of the canyon as we turned homeward.

The people were aloof and stand-offish I noticed and probably with good reason. There was also a good amount of the drug cartel moving into the area as well. We came face to face with four cartel soldiers in their just purchased, shiny new red Jeep Cherokee that was crossing the river. It was an odd thing to see, as we were traveling in the exact same red colored vehicle. We were on hyper-alert when we encountered these intimidating men. We did not want to provoke them in the least to become interested in our destination or us. These people are not above kidnapping and extortion. Luckily, they had other business to attend to it appeared as they waved us off as a minor interference while crossing the rugged rock strewn river bottom.

We explored the Spanish church at the bottom of the Canyon in a place called Batopilas that had giant church bells brought over from Spain in Spanish Galleons to Mexico and then carried inland in the 17th Century. It was hard to imagine. On the backs of beasts of burden and of suffering people, the ancestors of these natives became the Spanish slaves for the silver ore.

Something was beginning to change inside of me. I was falling in love with Mexico on a whole new level of experience. As a child in El Paso, I was afraid of seeing the suffering in people's eyes when I looked at them on the streets when my parents visited Juarez, Mexico. I was tender hearted and sensitive and Mexico frightened me. People were hungry and had so little. Here, in the Sierra Madre Mountains, was an intensity of physical deprivation of basic needs that I witnessed along with feeling strong emotions of compassion, empathy and I dare say, love for the land and her people so fiercely finding a way to survive.

It was going to take me many years from this meeting of the Tarahumaras to understand and link this physically powerful and mystical bloodline, or race of people to the giant crystal caves of Naica. I wonder if they knew of these crystals to be an important part of their territory in the rugged Tarahumara Sierra Madre?

CHAPTER 7

Oaxaca, Mexico and the Rancho Feliz Relief Fund
Winter 1997

I had just sat down on one of those fake Naugahyde barstools so reminiscent of bars of the 1940s in southern Arizona. We were sitting inside an old drinking establishment that had seen its better days, on the main street just north of the Mexican border-town of Agua Prieta, State of Sonora, and in the town of Douglas, Arizona. My friend Gil and I ordered a couple of cold beers and were silent for a few minutes. We attempted to adjust our eyes to the dark surroundings just after being outside in bright desert sunshine.

It had already been a busy weekend and it was only Saturday afternoon. I had flown in from San Francisco for a project and to meet up with Gil in Phoenix and drive down to Agua Prieta to deliver 8,000 pounds of canned goods to a food bank. We had just returned from visiting the soup kitchen that Rancho Feliz Charitable Foundation/Organization had helped to build for the less fortunate children that didn't have enough to eat during school hours.

After a couple of cold swallows to ease our thirst, we started discussing the poverty problems we had encountered on the other side of "the border".

Rancho Feliz had been providing food and shelter to many of the less fortunate in Mexico. It started in 1988 when I first came on board as a volunteer living in Phoenix, Arizona.

I had already been a mentor for a child in the Big Brothers/Big Sisters organization but it just didn't seem like it was enough. My little sister had problems that I tried to help with but I couldn't reach her. She had everything except the love and emotional caring she needed from her divorced parents. I tried to fill that with my time and attention but she continued to act shut down. I felt helpless in my ability to really care for her.

Growing up as a border kid of El Paso and Juarez, Mexico, I had seen plenty of humans subjected to the most terrible conditions of poverty, malnutrition, disease and physical handicaps.

Getting involved with Rancho Feliz's vision of helping the less fortunate was something I could take on with a deeper understanding I already knew in my bones. Besides, I could speak border Spanish and that would be a big help in the beginning of our endeavors and my future journeys into Mexico.

We had already been doing a tremendous amount to support and help our neighbors of Mexico for ten years; starting with efforts for the people of Agua Prieta, and further south of the Arizona border, the Tarahumara Indians of Copper Canyon and the pueblos surrounding the State of Oaxaca, south of Mexico City.

We ordered a couple more cold Tecates with lime, from the bartender who was overseeing a good paced business on a late Saturday afternoon. As we waited for our beers, a customer slipped a dollar bill into the jukebox and we listened to the Texas Tornadoes wail something in Spanish. The topic turned to southern Mexico.

Rancho Feliz established a relief fund in late 1996 in Oaxaca. We were buying food supplies and distributing to the pueblos nearby.

I started to feel a buzz of warmth and relaxation as I quenched my thirst with cold beer when my thoughts turned to Oaxaca and the stories I had heard from my good friend of his adventures there.

I wanted to do something more for the people in Mexico and the sense of adventure was tugging at me once again. Besides, I wanted to work on my almost native language of Spanish and hoped for fluency by being immersed in their daily life at the pueblos.

We agreed that I would go down there and check out what more could be done to help the sick, aged and poor people in the area. I would experience for myself the plights of the less fortunate of Oaxaca and discover a personal mystery of an ancient major center of

Mexico civilization, Monte Albán.

This ancient city, in ruins, was widely recognized by many people, many centuries before the Europeans arrived in 1500 A.D. But that is only part of the story.

Gil and I gulped our beers down and walked outside into a typical southwestern sunset full of neon pinks, oranges and reds. As we headed back to Scottsdale, Arizona, I was anxious to get started on this new adventure.

I left San Francisco on February 19th of that year, not knowing exactly what I was getting myself into. I had never travelled to Mexico by myself and was a little concerned about my safety. Little did I know I would fall completely in love with Oaxaca and her people, both the Mestizos and the Indigenes, as they call the native Indians.

There are currently sixteen distinct ethnic and linguistic groups in the State of Oaxaca, each with their own unique artistry such as pottery, textiles, tin and other handicrafts. The people live just like they did for hundreds of years, on terraced hillsides. Only now they live in tin shacks instead of the waddle and daub housing so typical of an earlier time.

One of the towns that I visited was Pueblo del Maestro. Theresa and her husband, Rafael, were my guides and Rancho Feliz contacts. One late afternoon, we drove out of the city and hit the dirt roads leading to the pueblo. As we drove up the terraced hillsides near sunset, the sky was lit up in deep oranges and blood red from all of the dust. Over to the west on the mountaintop, there was an outline of the pyramidal structures of Monte Albán, the first great city of Mesoamerica (500 B.C.) Imagining the powerful presence of the ancient aristocrats and nobles living in palaces and presiding over governmental affairs over the valleys of Oaxaca was such a paradox of the reality of what I was seeing right in front of my eyes.

My mind was full of questions. My heart was in my throat from such an awesome sight. I knew that I would return to Monte Albán as soon as possible to explore and learn what really happened there.

In the meantime, there was some important discovery work to do and report back to Rancho Feliz on the well being of the people we were looking after. There was a feeling of unease and I couldn't shake it off.

I met so many kind children, women and men. With so much poverty, lack of necessities and medical attention, it was hard to understand why these people always had time to give you a friendly smile. How can these people be this happy with so little? Growing up in El Paso, I had everything I ever needed. The feelings of gratefulness and sincere compassion started to seep into my heart. How unaware I had been of such levels of poverty. This was not a subject any of my friends in Scottsdale, Arizona and I were discussing. None of the people I knew of in the States were experiencing any of the sufferings of such a lack of everything. It was almost inhumane to witness the never ending flow of poverty and how it affected the people in this country.

After two weeks of visiting the pueblos of Oaxaca that Rancho Feliz supported it was finally coming to an end.

No doubt I was relieved but I was also weary of the heart tugging disconnect I would have to pull after feeling so welcomed and involved in a few people's lives.

Soon I would be leaving back to the U.S. but first, I wanted to make time to visit the huge central market and make my way to Monte Albán. I took the last tour bus in the late afternoon from the market in Oaxaca to the top of the mountain.

A brief history and description by Unesco says: "The historic center of Oaxaca and archaeological site of Monte Albán was inhabited over a period of 1,500 years by a succession of peoples – Olmec, Zapotecs and Mixtecs – the terraces, dams, canals, pyramids and artificial mounds of Monte Albán were literally carved out of the mountain and are the symbols of a sacred topography.

The grand Zapotec capital flourished for thirteen centuries, from the year 500 B.C to 850 A.D. when, for reasons that have not been established, its eventual abandonment began. The archaeological site is known for its unique dimensions, which exhibit the basic chronology and artistic style of the region, and for the remains of magnificent temples, ball court, tombs and bas-reliefs with hieroglyphic inscriptions. The main part of the ceremonial center which forms a 300 m esplanade running north south with a platform at either end was constructed during the Monte Albán II (*c.* 300 BC-AD 100) and the Monte Albán III phases. Phase II corresponds to the urbanization of the site and the domination of the environment by the construction of terraces on the sides of the hills, and the development of a system of dams and conduits. The final phases of Monte Albán IV and V were marked by the transformation of the sacred city into a fortified town. Monte Albán represents a civilization of knowledge, traditions and artistic expressions. Excellent planning is evidenced in the position of the line buildings erected north to south, harmonized with both empty spaces and volumes. It showcases the remarkable architectural design of the site in both Mesoamerica and worldwide urbanism.

Monte Albán is an outstanding example of a pre-Columbian ceremonial center in the middle zone of present-day Mexico, which was subjected to influences from the north - first from Teotihuacan, later the Aztecs - and from the south, the Maya. With its ball game court, magnificent temples, tombs and bas-reliefs with unexplainable hieroglyphic inscriptions, Monte Albán bears unique testimony to the successive civilizations occupying the region during the pre-Classic and Classic periods."

Try to imagine what it might have been like in the United States to see something so ancient with such an easy access. We have tremendous security and "do not touch" restrictions in place for many historic sites, with more than enough surveillance cameras. Now, being around all of this enormous history that was within reach made me dizzy with possibilities.

What could I find? Where could I go that normally would have been off-limits anywhere else in the world such as this place?

I made it to the Monte Albán museum where antiquities were encased behind glass that were said to rival King Tutankhamen's tomb in Egypt. I believed it after seeing jade, gold and other treasures. (Tutankhamen was an Egyptian pharaoh of the 18th dynasty; reigned c.1361– c.1352 B.C. English archaeologist Howard Carter discovered his tomb, which contained a wealth of rich and varied contents, virtually intact in 1922.)

The tour bus was honking its horn outside the museum calling the sightseers back as the site was closing its gates soon. It seemed I had just had arrived! I was determined to be back here in the morning for a full day of exploring.

That night, I went to the Zócalo (central park) in downtown Oaxaca to listen to music, drink cold beer and have a good meal. I couldn't believe how happy I felt. I was realizing that for years I hadn't experienced any community in city centers where families, friends, young and old would gather. That was in 1997. Today, the farmers markets are such a wonderful place for community to gather in any town or city.

I finished the last of my cold beer and slowly walked down the five hundred year old stone cobbled streets between the colonial casas and stately buildings. I listened to the lively music as I headed to my simple hotel room to sleep. The hotel was close by the five hundred year old, gold-filled Santo Domingo Catholic Church and I knew I wouldn't oversleep. The church bells start ringing around 6:00 a.m. every morning. With all the mescal (tequila) this city produces and consumes, I hadn't figured out yet if a headache was a result of those blasted bells or the alcohol.

I awoke to the sound of ringing bells in my ears but I was anxious to get started with my day to be annoyed. First stop was to see the Santo Domingo de Guzman Church.

As its name implies, the Dominican Order founded the church and monastery. Begun in 1575, they were both constructed over a period of 200 years, between the 16th and 18th centuries. The monastery was active from 1608 to 1857. In the period of the revolutionary wars, the buildings were turned over to military use, and from 1866 to 1902 they served as a barracks. The church was restored to religious use in 1938, but the monastery was made available to the Universidad Autónoma Benito Juárez de Oaxaca. In 1972 it became a regional museum, and in 1993 the decision was taken to undertake a full restoration. This was completed in 1999. It is an exceptional example of conservation architecture. The architect responsible was Juan Urquiaga.

The church has also been fully restored. Its highly decorated interior includes use of more than 60,000 sheets of 23.5-karat gold leaf. There was enough gold to convince the most skeptical of believers of the powers of the Vatican's far-reaching influence in the Americas.

The rooms that formerly constituted the monastery now house the Cultural Centre of Oaxaca, which was founded with the help of Oaxacan-born artist Francisco Toledo. This museum includes an important collection of pre-Columbian artifacts, among them the contents of Tomb 7 from the nearby Zapotec site of Monte Albán.

The former monastery garden is now an ethno-botanical garden, containing a large collection of plants native to the region.

The entrance to both church and museum is across a wide plaza that acts as a center for local fiestas and other entertainments. It is located about half a kilometer north of the central squares of the city, the Zócalo and the Alameda, and the connecting street is pedestrianized, so it is a popular place for both tourists and local residents to stroll.

I was in a hurry to get out on the street through the hotel's doors when I saw a most remarkable sight. There were two young boys holding the ankles a young girl in a school uniform of a green/blue plaid skirt with a white blouse hanging upside down. They were on the street corner of the great Church's property.

I hurried across the street to where they were standing, wondering what I could do to rescue the young girl from any harm. I was astonished to see all of them laughing and curiously asked them what they were doing? The strong smell of sewage had my eyes watering. Grasping for the right words to express my concern, in my poor use of Spanish, they responded in a pleasant tone. They explained that their friend (Amiga) had lost her gold necklace of Our Lady (The Virgin) of Guadalupe that had been hanging around her neck. It had fallen down into the ancient grate that was covering the sewage system below the street.

I could scarcely believe she was brave enough to endure the strong odors of sewage with her nose so close to the opened hole where the boys had moved the heavy steel grate to allow her head to be lowered into the dark passage below.

What was so enduring of this talisman that she would not only risk injury but embarrassment by being held upside down by two boys? Probably her parents had given it to her and would be very mad at her for losing it.

The Virgin of Guadalupe is the patron saint of Mexico. She is depicted with brown skin, an angel and moon at her feet and rays of sunlight that encircle her.

According to tradition, the Virgin Mary appeared to an indigenous man named Juan Diego on Dec. 9, 1531. The Virgin asked that a shrine in her name be built on the spot where she appeared, Tepeyac

Hill, which is now in a suburb of Mexico City. Juan Diego told the bishop about the apparition and request, but he didn't believe him and demanded a sign before he would approve construction of the church.

On Dec. 12, the Virgin reappeared to Juan Diego and ordered him to collect roses in his tilmátli, a kind of cloak. Juan took the roses to the bishop and when he opened his cloak, dozens of roses fell to the floor and revealed the image of the Virgin of Guadalupe imprinted on the inside. The tilmátli with the image is on display in the Basilica de Guadalupe in Mexico City.

The appearance of the Virgin of Guadalupe to an indigenous man is said to be one of the forces behind creating the Mexico that we know today: a blend of Spanish and native blood. Her dark skin and the fact that the story of her apparition was told in the indigenous language of Nahuatl and in Spanish are said to have helped convert the indigenous people of Mexico to Christianity at the time of the conquest. She is seen as having a blend of Aztec and Spanish heritage.

Her image has been used throughout Mexican history, not only as a religious icon but also as a sign of patriotism. Miguel Hidalgo used her image when he launched his revolt against the Spanish in 1810. She could be seen on the rebels' banners and their battle cry was "Long Live Our Lady of Guadalupe."

Emiliano Zapata also carried a banner of the Virgin of Guadalupe when he entered Mexico City in 1914.

Pope John Paul II canonized Juan Diego in 2002, making him the first indigenous American saint, and declared Our Lady of Guadalupe the patroness of the Americas.

I was guessing that this young student had already experienced the powers of protection of this strong female Saint and had no doubt she would do it again.

As I watched with total amazement, they lowered her body even further into the sewage passage. With unbelievable deliverance, she resurfaced carrying the gold necklace in her tiny hands.

This was a special sign for me. I knew something important was going to be experienced today. I felt it in my bones. I had to hurry now and return to the Mercado del Centro to catch that first tour bus to the top of Monte Albán. It was going to be a long day of exploration and I was looking forward to it.

It was already hot by the time I stepped out of the tour bus. I made my way through the entrance and into the museum and out through the back doors onto the site itself.

I wanted to start exploring the east sloping side where no one was showing any curiosity yet as the day was early and the doors to the site had just opened. I came across an entrance to the tomb called Number 7. It was locked up with a lock on iron bars over the opening of the passageway unfortunately, but I was able to see through the iron bars to view an exquisite carved statue on a stele.

I didn't read Spanish that well and mostly the important information in the museum was in Spanish with very little in English. I was not prepared emotionally for the torrent of information on this site that I would read about later. This was probably a blessing as this city had tortured and sacrificed many of its prisoners and indelibly etched these images in carvings of granite stone stele.

Tomb 7 represented a wealth of jewels and artifacts equal to what was found in Egypt's King Tutankhamen's tomb? Who was this deity, Goddess or aristocrat that was recognized by such wealth and stature at Monte Albán with over 200 tombs found below the mountainous Grand Plaza?

According to Tomb 7 at Monte Albán, it is one of the richest burial contexts ever encountered from pre-Columbian Mesoamerica. It was discovered by the great Mexican archaeologist Alfonso Caso in 1931, and it quickly became famous through popular publications such as National Geographic. In 1969 Caso published the definitive work on the tomb in his "El Tesoro de Monte Alban", which catalogued the hundreds of exotic objects of precious materials. Of particular interest, especially to Caso, were 34 carved bone objects decorated in Mixtec-style iconography.

Archaeological evidence at Monte Alban demonstrates spinning and weaving of textiles tools. There was a description of ceramic spindle whorls that had been found in Tomb 7. Additional artifacts from the tomb also correspond to spinning and weaving tools, including small spinning bowls made of onyx and crystal, and the carved bones that may have served as effigy weaving picks and battens.

Although previous interpretations of the tomb context had not featured an engendered perspective, A Dr. Geoffrey McCafferty was able to recognize a nearly complete 'weaving kit' . With this strong association with female tasks, the next step was to re-evaluate the context of the tomb itself, and the biological assessment of the sex of the individuals. Tomb 7 was first constructed in the Late Classic period, beneath the patio of an elite residence just north of the main acropolis. Several hundred years after its abandonment, the building was re-used as a temple. At that time the tomb was re-entered through the roof, and a new layer of deposition was placed over the original floor. This included nine skeletons, about 500 grave offerings, and a small altar upon which was placed a human skull covered with mosaic plaques of turquoise and shell. The majority of the skeletal remains were disturbed, although there was enough evidence of articulation to indicate that they had originally been primary interments. The most complete skeleton was identified as Individual A, and was in a flexed, seated position. The majority of the grave offerings were found in association with Individual A.

With the strong material evidence for a female gender identity, and the suggestive evidence that Individual A was misidentified and may actually have been female, Tomb 7 can now be reinterpreted. We suggest that it served as a shrine dedicated to a member of the earth/fertility complex. This tomb was sequentially re-visited by petitioners seeking oracular knowledge. This interpretation is informed by information in the Mixtec codices as well as other pre-Columbian and early Colonial pictorial manuscripts. Seven human mandibles were found in the tomb, painted red and perforated for suspension, in the vicinity of the mosaic skull on the altar.

This 're-spinning' of the symbolic significance of Tomb 7 places one of the richest offerings in Mesoamerican history within the sphere of female power, and was one of the first attempts to engender the ancient Mesoamerican past. It seemed as though an invisible force was guiding my journey into a weird synchronicity of events.

As I looked around for anything connected with this tomb, I started to sift through the stones, old pottery and dirt that had already been through a cleansing by teams of archeologists since the 1930's. I was looking towards the top of the tomb when I saw a

Mexican man wearing a white straw cowboy hat staring down at me. I was frightened only because I feared I had done something wrong. But it couldn't be further from the truth.

As I climbed up onto the Grand Plaza and started walking towards the south end where the cosmic observatory was, he started to follow me. He kept his distance but continued to follow me to every part of Monte Alban I visited. I finally came up to him and said hello and that began an extraordinary conversation.

If you have ever attempted to communicate with another person who spoke a different language than you, there requires a pathway that leads to understanding. For me, that understanding comes from emotions communicated from the heart, expressed through simple words. That seemed to help this man and I bridge a vast difference between us that was our ancestral histories and our lives.

I told him in such simple words in Spanish how honored I was to be at this extraordinary place. He agreed with a smile. Nervously, I introduced myself. He did the same. Then there was silence between us as I looked at this vast pyramid complex. I trekked over to the west side and looked downward towards farms and a village that were built on terrace. I was wary of his presence, as he had followed me. He then stood next to me as he told me that his family lived at that farm directly below us. He told me he was Zapoteca. I was astonished to learn that his family had been there for many generations and told him so.

I was hoping that he would move on, not understanding what he was doing there in my naivety.

He followed me around the entire day in the hot sun as I roamed the Grand Plaza with its crumbling and fallen stones, until the sun went behind the clouds in the late afternoon. It was nearing closing time again and I was beginning to get tired.

I was on my way out of the complex when he approached me directly and had asked if I would want to see something. My curiosity won and I agreed.

He had taken me to an area that was closed off to the public because the teams of archeologists were excavating this area. We ducted under the "do not enter" signs and ropes that held this site. He pulled back the corrugated tin that was covering the hole/passageway hoping no one would see him doing this. There was a ladder leading down and motioned me to climb downward.

I was fearful but not terrified he would hurt me. He saw my hesitancy and in my lack of understanding, he showed me what he wanted me to see. There in the dark shadows, was a vast column made of stone underneath on a dirt floor.

I was thinking while my heartbeat rapidly. All of the money I brought with me to Mexico was here with me including my passport and were inside my shirt. If there were ever a time that I felt vulnerable to robbery, this would be it. It seemed that some force was controlling my sensibilities, as I couldn't help myself from continuing to see where all of this would lead.

I showed him that I wanted him to go first down the ladder and he did. He helped me down by extending his hand and then went over to a very dark area underground where there was a shaft leading to another level down underneath. I was beginning to become really scared when he took a stone and dropped it down the shaft. We listened for what seemed like thirty seconds until the stone hit something. I was blown away by this event and felt like I was dreaming or was this was the beginning of a nightmare.

I thought without hesitation this was the moment and place in time he was going to throw me down this shaft if he could. I ran away from the darkness to where the ladder was and what light was left of the day reflecting down into this area. I thought for sure he was going to overtake me but he stayed put, near the passageway where he threw the stone.

I turned around when I realized he wasn't following me and said in a strong tone in Spanish: What do you want?

I sensed his hesitation as he struggled with the words. He walked towards me as he put his hand inside his shirt. I almost fainted thinking he was pulling a knife out. Instead, he brought out a newspaper wrapped object. I was shaking with fear and tried to control it. What he unwrapped totally blew my mind.

It was a graduated spherical green jade bead necklace that was ancient and simply gorgeous in size and color. I knew it was old by the encrusted dirt and pure color of jade. I was simply stunned to witness this as he place the beads on the stone bench-like structure underneath the Grand Plaza. How in the world did he acquire this antiquity? My mind was racing attempting to find a deeper understanding of the situation.

In a heart felt way, he told me what I felt was the truth. His family had been part of the workers who were excavating the site and they had stolen this piece from inside the chest of a skeleton buried in one of tombs. (There are 200 tombs that are underneath Monte Alban.) How many other items were taken from this site by the workers? He knew it was an important "piece" and he wanted to sell it.

I had seen all of the jade jewelry in the Museums and knew this looked identical. What I didn't understand was how did Mexico acquire jade? It was beyond my knowledge at the time.

Here is what I discovered years later: Scientists Solve Jade Source Mystery - June 2002, World Scientist:

"Since the 18th century, collectors, geologists and archaeologists have sought the answer to a frustrating mystery: The ancient Olmec fashioned statues out of striking blue-green jade, but the stone itself was nowhere to be found in the Americas. Now scientists believe they have discovered the source, a mother lode of jade in Guatemala that could tell much about ancient American civilizations and about the formation of the continent where they lived. Ever since Alexander von Humboldt began collecting jade in Latin America in the 18th century, Olmec-style statuettes and axes, crafted more than two millennia ago, had been found from Mexico to Costa Rica. But never had that kind of jade been seen naturally in any quantity in the area. Then in 1999, Russell Seitz, a geophysicist who had spent 23 years searching for the source of Olmec jade, visited the colonial city of Antigua in central Guatemala. On the roof of a store, he found jade that was vastly different from the opaque jade he had seen in Mexico and Central America, and it was identical to the translucent blue-green stones so coveted by the Olmec, who lived in central and southern Mexico from 1000-400 B.C."

Could this be the source of these gorgeous beads? I am sure it is.

I then asked this man why he wanted to sell this exquisite necklace to me? He had told me that his family was poor. There were extended family members and many needed medical care.

I knew this to be true of the many pueblos I had visited. Yet still, why me? I certainly did not look like a person of wealth by any stretch to afford something like this. Also, it would be foolish for me to take this necklace back to the states and across the border. I would go to jail if this were caught in my possession.

He pleaded and bargained with me to take the necklace and I continued to refuse him. That is when I knew it was time to scramble up that wooden ladder quickly. Would he try to pull me down? I really didn't know the answer to that and made sure I made it up to the top of the Plaza as fast as possible.

Again, he didn't follow me. He waited then took his time to climb up as well. That surprised me and for some reason, I didn't run away.

He looked straight into my eyes, his as dark as night, with a black mustache on a very brown face. He said in Spanish: "Who are you?" "Where did you come from?"

I was surprised at this and dared to respond truthfully to him. I told him about my healing work that I did in California and my family story from El Paso, Texas.

As we conversed in very simple Spanish, the sun was beginning to set. We obviously weren't through with our communication. We decided to hide from the public and sit behind a mud and stonewall near the entrance and continued our conversation.

Something happened that was so inextricably hard to single out to describe. I felt comfortable with him, like I had known him. You could tell he did too. There were many moments of silence since we barely could talk with each other in my poor understanding of conversant Spanish. The air was charged with energy as we dared at times to look each other in the eyes.

When the words did come, they were powerful because they became words expressed from the heart. He told me he had never met anyone like me before. I was feeling something akin to remembrance of a dream, or another time, as I received his words. He was grateful for the inexplicable introduction that synchronicity had brought to us. I humbly accepted his expression and thanked him from my heart to his.

The time was nearing an end now. The gates were closing and the tour bus was waiting for an accounting of all passengers. We stood up and he placed his hand out to mine. I was so moved by this that I touched his shoulder and placed my hand in his outstretched hand. There was an emotion in his black eyes and there were tears welling up in mine.

This moment felt bizarre and I could hardly contain my emotions.

The last thing I read about Monte Alban was that it had only been less than ten percent excavated in 1997.

As I boarded the bus, I turned towards the Great Plaza where he was standing staring at me. I waved goodbye.

I sat in a stunned silence as we trundled down the mountain and wondered about the series of strange events that had just happened. What other treasures would be found and revealed in this mysterious place of the early Olmec?

I never returned to this place.

CHAPTER 8

The Grand Canyon and the Longest Foot Race
October 1997

It was the early fall in 1997 when the Run Across Arizona extreme sporting event and the Rancho Feliz fundraiser began.

I had been involved with Rancho Feliz since 1988. We had created some really off-the-wall and fun events for people to donate money and goods to the less fortunate in the border town of Agua Prieta in Sonora, Mexico and the Tarahumara Indians in Copper Canyon, Mexico. And to this day, this organization is still alive and well and we are continuing to raise money each year for various projects for the citizens of Agua Prieta. (See: www.ranchofeliz.org)

My good friend, Gil Gillenwater, founder and visionary for these crazy, almost to the breaking point type of endurance events, would draw the interest of fanatics who loved extreme sports. With Arizona's dry and sunny climate, this was a breeding ground for them.

I was more interested in the periphery of these off the wall events acting as infrastructure and support for the participants whom I secretly questioned their mental stability.

This event was going to be more complex to pull off than the other events we had done in the past. We brought in the ultra long distant runners from Copper Canyon, the Tarahumara Indians, to run with our relay group.

The Run Across Arizona had 10 volunteers and 4 Tarahumara Indian runners who ran a 678-mile, 114-hour "non stop" relay from Kanab, Utah to Agua Prieta, Mexico. Donors were encouraged to pledge for each mile ran and in the end $160,000 was raised from this event for the charitable organization, Rancho Feliz and for the Tarahumara tribes that were living in Copper Canyon.

My job as part of the support team was to be a scout using my vehicle to assess the next leg of the relay race and to make sure there were no unanticipated dangers that could potentially harm the runners. This was an ideal volunteer job as there was nothing I liked better than to be traveling the back-roads of Arizona, one of the most picturesque states in the U.S.

As we were gathering near the Utah border to kick off this rather grueling run, a long blue and silver van drove up. Out of the slide door, the first of four Tarahumaras stepped or jumped out. All were wearing the traditional apparel and white straw cowboy hats. Quiet and super observant, these men stayed very close to one another. I am not sure if any of them spoke English or just Spanish and their home language of the Rarámuris.

This is when I met Richard Fisher, their running coach and sponsor/manager as he turned off the engine, kicked his door open and stepped out of the van. There was an excitement in the air, as this team was already well known in the global running world.

One such example was the great Native American Champion, Carrildo Chacarito. He took first place, age 43, in the Angeles Crest Trail 100 mile Race, Sept 27, 1997 in Los Angeles. Carrildo was second place finisher overall at the Leadville 100 miler in 1993 in Colorado and at last was able to prove his championship abilities in 1997. He was the last Tarahumara international champion.

We were going to begin almost 5 days of non-stop relay running across Arizona that included approaching the north rim of the Grand Canyon and running down the North Kaibab trail to Phantom Ranch, 13.6 miles, that is located at the bottom of the canyon next to the Colorado River, then run the trail back up to the south rim onto the Bright Angel Trail that would be another grueling 9.9 miles.

This running the canyon's rim to rim part of the race would be the toughest portion across Arizona, as there would be no possible way to offer a support team if there were any injury or harm to the runners. And besides, who would be able to keep up with the world's fastest runners, especially in the Grand Canyon? Except for one guy, our dear friend, Bob Kite from Arizona who did run with them. I wish I had talked with him more about that experience.

We knew that this would be where the Tarahumara would excel. Copper Canyon is more than twice as deep as the Grand Canyon is at 5,300 feet in elevation - however no one of this running team had ever seen the Grand Canyon before or run the rim to rim.

A storm was setting in and the wind brought ice with snow and rain swirling around us when we approached the north rim of the canyon nearly two days into the event. I was feeling nervous and excited and cold as hell as I watched these men in their little shorts and sandals not even twitch from the chill and ice hitting them in the face.

I remember parking my little silver bullet of a scout vehicle and running to where the team was standing on the edge of the rim. I wanted to see the look on their faces, as I am sure they were the first of their village if not the entire elusive Copper Canyon dwellers who had witness such an awesome view of our Grand Canyon.

Their faces were serene as if there was a feeling of being at home and relaxed in this rugged country although their sandaled feet had never walked here before. I watched them observe everything with a level of scrutiny while they smoked cigarettes. This was a first in history and I was privileged to witness it.

I was humbled and feeling somewhat guilty for what these runners were going to endure over the next several hours - from snow, sleet, rain and wind at the higher elevations to temperatures that could be over 90 degrees at the bottom of the Canyon in early October.

All four of them were going to have to run together the entire way as it was not possible for the other 10 runners to keep up with them, nor would they want to. At the top of the south rim near the trailhead, another runner would be poised and ready to take the baton and begin the next leg of the race. An easy effort compared to what went on before.

It was only later that I found out that these fantastic runners believed they were going to have to run the entire distance of 678 miles, every single mile of it from border to border and were prepared to do it! And you can believe they would have done it. No question about it. Imagine the relief they must have felt when it was explained to them they only had to do the most grueling part of the race and to our astonishment completed the rim to rim of twenty-three and one-half miles (23.5) in less than six hours and then continue to run every fourteenth leg of the entire journey until they reached the Mexican border in southern Arizona.

The rest of us had a long five-hour and 214 mile drive to get from the north rim to the trailhead of the Bright Angel Trail on the south rim to deliver the next runner awaiting to take the baton. We were the lucky ones.

I did some praying that day. I had hiked the Grand Canyon from the south rim side down to Phantom Ranch and returned to the top in one day several times already in my life. It was strenuous and enthralling to be able to push the human body to that extent.

It made for long days beginning before daybreak and finishing well after the sun had set down. And if someone were to be injured down at the bottom, it was not easy to get an air evacuation out of there. This was not a place to be careless for no one would be coming to rescue anybody for a very long time.

Rick, the Tarahumara running coach, was temperamental and overly protective. There were a lot of misgivings as to the safety that he could not guarantee for his runners. I did my best to either stay out of the way or offer solutions that could be heard. Mostly, I kept silent but maintained a positive and energetic attitude.

I had been mystified with the reclusive Tarahumaras when I was in Copper Canyon two years earlier and found them to be almost magical in their superhuman ability to run ultra-long distances.

After 114 hours later, with friends and loved ones and crew-members, we watched all fourteen runners run together and cross over the finish line on the border touching the well known fence between our two countries of the U.S. and Mexico near Douglas, Arizona.

We had a very happy celebration at the Gadsen Hotel in Douglas where there was a stampede of thirsty runners and crew alike eager to begin drinking several rounds of beers.

Logging over 3,000 miles on my car traveling north and south on the back-roads of Arizona, it had been one exciting journey. All in all, it was an amazing time of tears, laughter; sleep deprivation, weather extremes and tremendous satisfaction for raising funds and awareness for the poverty-stricken and less fortunate beings living in Agua Prieta and the Tarahumaras in Mexico through the Rancho Feliz Organization.

It was only after I had entered the crystal caves in 2001 that I was to learn that the town of Naica and the Naica Mountain (founded in 1828) was a Rarámuris word named by the Tarahumara for the location where the giant selenite crystals were going to be found in 2000: A place of shadow - or other translations say it is a waterless place.

A paradox between name and place, when the Spanish took that term from the Tarahumara language, I am sure they didn't know that what lay beneath their feet was one of the world's largest aquifers that gave birth to the mysterious giant crystals.

CHAPTER 9

One of the Worlds Greatest Natural Marvels
April 2000

The Discovery News Channel: *"April of 2000 - Giant crystals are found in caves 1,000 feet below the earth's surface in Naica, Chihuahua, Mexico".* This announcement in the news was the beginning of how my extraordinary exploration into the giant selenite crystal caves of Naica, Chihuahua, Mexico begins.

The story starts to take shape while I was in Agua Prieta, Mexico on October 20, of 2000, once again fund raising for the Rancho Feliz Organization and giving out early Christmas toys to the children of the orphanage.

There was Rick, the coach and sponsor of the Tarahumara running/racing team in 1997 and one of the most widely known canyon explorers in the world, quickly leaving the grand lobby of the Gadsden Hotel at the U.S. border town of Douglas, Arizona.

It was a very cold day in the early fall that year with blustery winds that pulled dust devils through the streets and sand drilled down the sidewalks. As I parked my car after a long day's journey from California, I recalled from other Rancho Feliz events, of seeing Mexican children playing in the dirt roads without shoes at 23 degrees Fahrenheit and was astonished to witness this. I shivered as I pulled my jacket around my neck and ran towards the hotel entrance.

As I was pushing the glass and high polished brass front doors open into the lobby, Rick was walking out. We came face to face at the doors and once again greeted each other after three years since our last meeting at the Run Across Arizona fundraiser for the Tarahumaras.

"Where have you been?" he asked, as I stood there stunned by his question with no quick reply to the inquiry. "I've been looking for you for the last three years and I haven't known how to contact you. Listen, could you come with me for a moment or two? I want to show you something."

He asked me to step outside and walk with him to his van that was waiting and parked on the main street. I was so surprised with running into him again that I followed without hesitation and with no reply.

He had some exciting news and wanted to show me something. As he leaned forward into the van and reached for his briefcase, I started to shiver again from the impending rainstorm looming in the distance. He pulls out six color photos of some very large crystals.

What I was looking at resembled what Superman's Fortress of Solitude must have looked like on a distant planet: massive, heavy crystal structures as thick as pylons and resembling huge beams growing at every angle wide enough and large enough to walk on as planks.

My mind could not even begin to register the immensity of something like this. My eyes were transfixed and I was having difficulty determining the size as my mind wanted to play tricks with me as to the scope and dimension. I was imagining little miniatures that were blown up to larger than life sizes to look large but were very tiny. Something a Hollywood movie studio might have created in a movie-set (as they eventually did for a movie entitled "The Core starring Hilary Swank").

I asked him what this was? He looked at me in total amazement as if I hadn't understood what he was attempting to show me. I hadn't wanted to appear stupid but confessed I didn't know what the hell he was showing me. He had to explain to me that these were the largest crystals found on the planet and they were in Mexico. What was more fascinating was that he had shot these photos himself.

I started to get very excited when the information started to register in my brain. When I looked at the images closer I started shouting with excitement. I couldn't contain myself. I had never seen anything like this and I could hardly believe what I was seeing. Large greyish blue translucent giant crystals that hung largely from the roof and coming out from the sides of the cavern at angles that seemed to defy gravity. Was I in a dream?

Rick laughed at my enthusiasm and told me to hold on to the photos. He would ask for them back at a later time. But for now, he needed to get on the road immediately. He had little time to visit but wanted to stop in Douglas to see our mutual friend. He was on his way to the State of Chihuahua, Mexico, with an invitation from the Penoles Mine where the crystals were and an escorted group of Mexican government officials.

Being the first American to photograph these unheard of giant crystals and to do it properly required his expertise since there was great difficulty getting any clear images out of the bubbles inside bedrock in the mines. I didn't have time to ask why clear images were so difficult to obtain at the moment. The question was registering deeply in my mind for a later discussion with him.

Knowing that Rick was an explorer, I was sure that what he was up to was a spectacular find that no one on earth knew about except for the people at the Penoles Mining Group in Naica and a few people at the Dept. of Tourism in Chihuahua. As usual, he was in a hurry to get on the road to Naica, Mexico and made me promise to return the photos to him at a later date.

Months later, after I had relocated to Sedona, Arizona, for a brief time, I finally make an effort to contact him. I was living in the metaphysical capital of the world known for crystals, energy vortexes and healing with crystals and I had been sitting on those photographs for two months, not telling a soul about these amazing photos I had in my possession. I don't know why I didn't say anything but it felt like this was something sacred. It wasn't until February of 2001, that I learned that Discovery News had published an article about these crystals in April of 2000.

It was January 19th, 2001, when I phoned Rick to inform him I was reluctantly returning his earliest photos of the crystal caves back to him

He started laughing and said, "You have called in the nick of time. I am leaving Monday morning at 6:00 a.m. for Chihuahua City on another government expedition to the mines in Naica to explore the two new cavities of crystals only discovered in April of 2000. There is a space available for you if you want to come." he said.

Before I could even let out a gasp of surprise, or respond, he said: "However, I am photographing the crystals not only for the government but for an outdoor clothing company who is sponsoring this exploration. This clothing is somewhat similar to the type as Patagonia. The catalog offers clothing for extreme climate conditions. The name of the catalog is Rail Riders. Have you heard of it?" he asked.

"Oh yes, I have heard of it," I said. "They make a special kind of clothing that is used in tropical and other extreme temperatures where the humidity and heat are very high. Normal clothing such as cotton deteriorates on the body in a very short time in this type of climate. This fabric dries in a matter of minutes, even in extreme humidity." I said. I had remembered seeing the catalog in previous months. There were sidebars on several pages of the catalog depicting people's adventures doing amazing things while wearing the manufacturer's clothing and so had peaked my interest!

"That's right, and where we are going is a perfect example of that type of climate." Rick stated in a definitive tone of voice. The palms of my hands were starting to sweat ever so slightly. I had a feeling that I was going to receive the answer to that long held question as to why the photos were so difficult to shoot in the caverns.

"Listen", he said in a very serious tone and then paused, listening to me on the other end of the phone, making sure I was paying close attention to the following explanation he was planning to give me. He wanted to be as clear as possible with his next words: "In these extraordinary crystal cavities, there are volcanic fissures in the bedrock that are heated by a chamber of magma one mile below the underground aquifer With no natural entrance, these are cavities inside the bedrock that are busted open by heavy-duty mining machinery."

"This bubble inside bedrock is what has helped to grow the crystals to such gigantic proportions. The cavities are closer to 100% percent humidity right now and slowly dropping. The smaller cave is approximately 128 degrees in temperature and the Cave of Giants (later named the Cave of Crystals) would likely be the same temperature and both are slowing dropping in temperature. It is a perfect condition to experience the durability of the clothing in extreme environments. Are you still interested?" Rick asked. I was starting to feel this queasy feeling in my stomach. "Yes" I said with false confidence as I swallowed a hard lump in my throat.

Rick started to strategize his intentions with me and said: "What I wanted to demonstrate and photograph on this expedition is a more feminine quality since I mostly have only miners or engineers in all my photos with the crystals. There is no natural light of course, just the small light on top of a miner's helmet. It is a black hole inside those cavities, with a hot and steamy sauna-like environment."

I'm thinking oh, that's just great. Then it dawned on me about my other encounter with crystals and black holes back in 1998 in Plumas County. What kind of parallel of different timing was this?

I needed some time to think his invitation through. I told him I would consider it and would call back later. Rick warned: "Don't take too long to get back to me because we are leaving for Mexico day after next in the early morning."

It took several hours for all of this to sink in. I couldn't believe it, but I was actually thinking of not going because of other commitments I had in Sedona. Like that new job I was just hired for.

Then it hit me like a ton of crystals fell on my head: I was probably going to be one of the first American women to explore these newly discovered caves. That is if I had enough courage to overcome my fear of black holes and claustrophobic, scary places beneath the earth, now known as one of the most hostile environments on earth.

Well it didn't take too long to make a decision. Even if it meant not having a new job in a new town I was living in. But I will confess, I had a good cry out of fear because I was scared. Scared of being lost or being trapped down there inside that mine.

But when was an opportunity like this one ever going to present itself again? I knew I had to go and step out of my comfort zone on all levels to experience this once in a lifetime adventure and exploration.

I called him back later that afternoon. The next thing I knew, my bags were packed and I drove off to Tucson to meet Richard that Sunday night.

I was thinking while I drove the 5-hour desert drive from Sedona to Tucson; well at least I wouldn't be cold in the caves. Besides it was wintertime and a warmer temperature would be a nice change, I thought while a little smile was developing on my face. I was driving south into Tucson, heading further into an unknown adventure that had dangerous implications.

CHAPTER 10

The Sierra Tarahumara Occidental and Her Treasures
Winter, 2001

It was early the next morning, on January 22nd, 2001, that we left Tucson and drove sixteen hours straight to arrive at the Palace of the Sun Hotel in downtown Chihuahua City, compliments of the state government. The city was located 150 miles south of the border, from my hometown of El Paso, Texas

For driving in an endless landscape of desert scenery, it was an interesting road trip where there were many tales of adventures shared with each other. Passing the dry Chihuahua desert in the winter sun, Rick told us about many extraordinary journeys he had taken all over the world including working closely with the legendary Tarahumara Indians from Copper Canyon, Mexico. Also, he told us of his incredible discovery of the deepest canyon in the world in Tibet. While we traveled the main highway south of El Paso, Texas and deeper into Mexico, I was preparing myself psychologically for what was going to happen the next day. I was scared. Scared of the unknown and I didn't want to admit it to him.

We were tired and needed rest by the time we pulled into Chihuahua. That night I dreamed of a white owl leading our expedition group to the mine and into the crystal caves. This beautiful creature spoke to me and said with excitement, "Hurry! Hurry! Come! Come!" She flew into the caves and invited us all to join her! She circled us and spoke to us although I could not recall what she said to me. Somehow my dreaming brought me a lot of comfort and a mild feeling of confidence.

I woke up with a sound intuitive forecast, that it was going to be a very auspicious day and for some deep reason that I could not logically explain, our group was to be protected from any harm or injury.

It was sunny but very chilly at 7:30 am as we stood in the shadows of one of the tallest buildings in Chihuahua. It was January 23rd, and we were waiting for the rest of the team from the State Department to show up and caravan out to the mining operation in Naica where the mysterious crystals were waiting to be fully explored.

We drove in a couple of vehicles southeastward through the small town of Delicias that was located on the famous Royal Road or the El Camino Real de Tierra Adentro (Spanish for "The Royal Road of the Interior Land"). This was a 1,600 mile (2560 kilometer) long trade route between Mexico City and San Juan Pueblo, New Mexico, from 1598 to 1882.

The trail was unofficially used for trade among native tribes (The Tarahumaras and others) since the earliest of times. It didn't become an official trade route until 1598 when Oñate followed the trail while leading a group of settlers during the era of the Spanish conquest. The duration of the trip from the Rio Grande to the San Juan Pueblo was said to take, by wagon and by foot, approximately 6 months including 2-3 weeks of rest throughout the trip. According to journals kept by settlers they used common animals found along the trail to add to the food they brought along. The trail greatly improved trade among Spanish villages and helped the Spanish conquistadors spread Christianity throughout the conquered lands.

The trail was used from 1598 through 1881 when the railroad replaced the need for wagons. Eventually, railroads replaced rutted trails and over time the trail and evidence of it faded from sight and memory. The changes that the railways brought made trade along El Camino much easier and in some cases made travel quite luxurious.

When we turned off the El Camino Real and drove west towards the Naica Mountains, it was the last stretch of road into Naica.

As we traveled closer, there were what seemed to be military soldiers standing in the road blocking our access. On closer look, I was not sure the soldiers were carrying double straps of ammo across their camouflaged green and tan shirts.

It looked as though we were back in the days of Poncho Villa the Mexican Revolution as I peered at their brown faces with army caps on their heads. I was suspicious that these armed "military" men were waiting for some payoff before letting us through. Living on the Mexican border growing up, I had seen plenty of trouble and delays with "Mexican officials". I was praying there would be no trouble and sure enough, after inspecting our vehicles in detail, they let us pass through.

In 1794 Messrs. Alejo Hernandez, Vicente Ruiz and Pedro Ramos of Verea located a small vein of silver at the foot of a mountain range known in short, Naica, south of the current city of Chihuahua. On June 6 of that year, they formalized its discovery by making the legal announcement that "a mine located in wilderness, with the name of San Jose del Sacramento in Glen Aguaje, of the Naica Mountain".

Previously, nothing was known of Sierra de Tarahumara (Naica). But my guess is the Tarahumara not only knew all about these mountains or mountain range to the east of the Sierra Madres but traveled the El Camino Real constantly and probably inhabited sites along the route.

Apparently the name Naica in the language of the Rarámuris (or the Tarahumara Indians) means "shady" nai, "place", and ka, "shadow". Other sources translate it as "waterless place". It would not be a far stretch for one to come to the realization that if the name of Naica was taken from their language to name this location, and that it was to be the most producing silver mine in Mexico in 200 years, then there had to be a strong connection to the Tarahumaras, whom I found just as mystical as the giant selenite crystals. However, at this point, I have not been able to do any further research in this area of connecting the history of the Tarahumara to the Giant Selenite Crystal Caves or if they were aware of the crystals before any miners discovered them.

Although there was a claim or announcement of its minerals, the first digging was not to be performed until 1828, when the town of Naica was established. Things went very slowly.

In 1896 Mr. Santiago Stoppelli then made a formalized claim of a mine in Naica Mountain. They soon formed the Naica Mining Company, and exploitation on a large scale began in 1900.

The importance of Naica was such, that by 1911 it had reached a municipality. However, due to the ravages of the Mexican Revolution, the company was forced to suspend mining, which was not continued until 13 years later by the Peñoles Mining Company, which went into operation for another four years.

Between 1928 and 1961 the Naica mine was exploited by U.S. companies. From then on, Peñoles operates the mine with great success, being one of the most important and productive in the state of Chihuahua and in Mexico. Currently the mine mainly produces lead, zinc, copper, and silver, processing almost one million tons of ore annually. The mine workings have been highlighted nationally for its environmental stewardship and minimum pollution.

It was mid morning on Tuesday, January 23rd, before we could gain access to the mine by the Penoles Company officials and security. Our instruction were to travel into the tunnels that would lead us 900 feet below the surface to find the two selenite crystal caves. We met up with the rest of our team: Carlos Lascanos, top cave explorer of Mexico, and Sonia Morales from the State Department and were greeted by the manager of the mine, Roberto Gonzales and Chief of Security, Alejandri Enrique Escoto.

We each were equipped with a mining helmet and battery pack belt. And that was it. Even though it was chilly on the surface, we all changed into the extreme clothing that Rick provided knowing we would be entering a zone of much hotter temperatures.

I wore the Rail Riders sleeveless white shirt underneath a marine blue long-sleeved shirt and shorts, a pair of tennis shoes, and knee protectors that never did stay up, because of the copious amounts of sweat that would shortly be coming out of the pores of my skin from the humidity and the heat deeper down inside the mine.

This mine is considered the most hostile environment in the world for mining personnel. As work in the mine would be around 102 degrees in the mining tunnels, much more hotter temperatures would be found in the cavities busted out of bedrock by machining for ore excavation. Also, at the far depths of the mine, over 2,200 feet below the earth's surface, the heat was even more intense.

We jumped into the Chief of Security, Alejandri's heavy-duty pickup truck and slowly made our way down one thousand feet. I did my best to keep my anxiety levels down by distracting myself with listening to the others conversing in the cab.

The heat started to increase and I sensed the darkness begin to surround us. I thought this was similar to what dreaming was. I closed my eyes and let the darkness and humidity envelop me.

What seemed about an hour later, we drove to where we could park the truck in the middle of the tunnel's passageway. We jumped out of the truck and stood at a place where lights were strung across the top of the tunnel's passage. There was an earthen pile of dirt and crystals bulldozed in front of the hole made by the miners to cover up the smaller cave's easy entrance.

I thought it was odd the mining company would do this but I was soon to find out why.

Afterwards, when we had resurfaced, the cavity with the smaller crystals was named the Cave of Dreams (later to be renamed by the National Geographic team members and the Naica Project as the Eye of the Queen).

We scrambled over the dirt and broken selenite crystals that lay all over the ground and walked up to a small wooden ladder that had been placed there for our exploration.

First images that were shot surprisingly would show the luminosity of the crystals lighting up with the flash of a camera bulb as if the entire place was alight by electricity.

We climbed up the ladder with only the low light from the overhead strung mining lights in the tunnel to help us see the steps. I turned my helmet light on and saw my way to climb through a hole and crawl onto jagged and pointed giant selenite crystals.

My body was hit by a blast of black and humid heat. That smell! It filled my nostrils. It was ancient dirt baking in a very hot furnace. The smell reminded me of what I imagined an ancient sarcophagus in a pyramid's tomb would smell like – except for the heat.

I crawled slowly forward into this dark cave with only a small beam of light, holding on to pointed tips of hot crystals with ungloved hands. I squeezed through a small cavity on my back carefully then popped into a bigger chamber.

What I was looking at was hardly believable. With steam still rising from everywhere I looked, I had to adjust my eyes by blinking the humidity away. This heat could scorch the eyeballs. What I saw looked like a huge blue crystal waterfall of flat rhombohedra shaped selenite crystals stacked and appeared frozen twenty feet high.

My miner's helmet did not fit right on my head. It fell forward onto my face from the weight of the overhead light attached and loose straps. It became a constant source of irritation for me, as I needed to pay attention to more important things such as where I was going to climb next. It was a struggle to stay grounded as I was beginning to lose all sense of myself here.

This little aggravation was going to be a blessing in disguise when only a few images were to be salvaged for academic and research purposes. Most of the images or slides taken were useless as a result of my miner's helmet's light beaming an angle right into the camera lens. As a result, the photographer was somewhat blinded by the light as he took shots. Later, upon examination, the photos would be tossed into a garbage can, determined to be of poor quality. I had asked the photographer if he wouldn't mind if I took a few out of the garbage bin for my own personal use? He granted me permission and that would turn out to be a most fortunate destiny for me.

I had no foreknowledge that these tossed slides of the photos would be the start of a new chapter or passionate purpose in my life. I would soon begin sharing these images and my adventure/story about the giant crystals with people from all over the world. It was going to take years of learning and research to speak clearly about these crystals and their impact. I had my work cut out for me.

CHAPTER 11

Frozen Crystal Water Fall – Cave of Dreams

The intense heat and enormous size of the crystals was overwhelming for me. How could I be seeing a frozen crystal waterfall that appeared to be cold as ice and existing in this hot temperature? I had to struggle with the logic of this visual, as all I had to see into this black cavern was the tiny light of my helmet.

As I looked across the chasm, I wanted to leap to where the crystal fall began but that was not possible. A chamber down below, between where we entered and where we were to climb next, separated it. We climbed down the twenty feet to the bottom of the chamber and then started up the other side onto the crystal waterfall. It seemed as though I could just walk up, but my sweating hands made the crystals very slippery and dangerous. The heat was starting to place a chokehold on my throat. My kneepads weren't doing me any benefit as they were already sliding down my shins from the moisture and my sweat from the heat.

Marveling over this incredible sight, I started to strategize my next moves – a gigantic crystal waterfall of selenite stood 20 feet high before me; onto which I had to climb or rather crawl up. The crystals were so beautiful and translucent. There was a slight hue of pale blue to them that seemed to reflect the blue clothing that I was wearing. When I reached the top of the crystal fall, the heat was even more intense.

My heart was beating so fast I thought I would not make it. This sensation wasn't just from the exhilaration of seeing something that my eyes could not actually believe was real. My organs were beginning to heat up from the inside with no way to cool off. Thirty minutes inside this cavern would kill you from the poaching of your organs. I know we were in there longer than thirty minutes and attempting to explain we had endured this environment, well, would be unexplainable and possibly a miracle.

But then again, there was that dream of the White Owl promising our safety the night before and for some reason this brought great comfort to me. I trusted this dream and I am sure this had a great affect on me psychologically.

The heat and sweat was pouring off my body and made it difficult to hang onto the crystals. The heat from the crystals was almost hot to the touch and the sharpness of the splintering effect of the selenite crystals beginning to dry out made it hard to concentrate where to look and where to move next.

It took all that I had not to panic. Behind the crystal fall were even bigger jagged selenite crystals that led to what seemed to be the back of the cave. It was pitch black in there. I was feeling the intensity of my fear. It felt as if we were inside a giant womb in Mother Earth. And indeed we were.

It is estimated at that time in 2001, that the size of this smaller cave was around 2,000 feet. I wandered into an area that had not been explored before while Rick shot in rapid fire, a multitude of photos. It was rough going and painfully slow and a race against time between exploring this unknown part of the cave and getting out of there before going unconscious and/or injuring ourselves.

There was one beautiful huge plank and/or pylon at the far end of the cave that had not been previously explored. It was my objective to get on that beam and to move beyond it. It was massive! With four solid sides of pure crystal, it had the feeling of a crystal bridge. All the pylons existing in both caves weighed an average weight of 60 tons.

I was becoming extremely anxious and did my best to not reveal it. Feeling dizzy, I was concerned I would fall down and really injure myself in the process. I thought Rick underestimated the heat. It felt like 130 degrees inside here.

One extraordinary selenite crystal seemed to be hanging in the air and was the size of a compact car. It literally seemed to defy gravity. The geometry of it was extraordinary. Rhombohedra lines that came to a perfect point resembling an upside down triangle poised right above my head. Many of the crystals were nearly transparent and beyond perfection. To think that no one had been here before was something I was to contemplate for many years to come.

Light was emanating through the crystals and reflected light back to me from every direction. This was happening from the light of the camera's flash bulb and from the miner's helmet into the faces of the selenite crystals. But there seemed to be one other light source coming through in areas that were not so lit up. Where was the light coming from?

Was I losing my mind or was I hearing a very low audible humming sound? I thought it might have been mining machinery but there was nothing that could have sounded that low in hertz cycles. Was it coming from the crystals themselves? I wondered if anyone else could hear that extraordinary sound? No one mentioned it and it was long after this experience that I would remember hearing it.

In attempting to explain this sound, I would say it sounded like earth herself was pulsing the Schumann Resonance (see: www.earthbreathing.uk.co/sr.htm) and I could scarcely believe that I may have actually heard it within the earth.

Attempting to solve the humidity affecting the cameras to get good clear images was extremely difficult but Rick had come up with a solution. A professional photographer with credits from all over the world, he solved the problems affecting the images that produced steamy or fogged photos.

The answer came from using clear glass filters over the lens and allowing the cameras to adjust to the extreme heat. No digital camera could do the job, only the manual ones that shot 35-millimeter film. Even then there could be and was a malfunction and so multiple cameras were brought into the environment for acclimation prior to use.

It was only in 2006 that the Naica Project brought in digital cameras wrapped in plastic that were allowed to acclimate to the temperatures of the caves for two hours prior to shooting. Even then there was no guarantee the cameras would not break down and quite often, they did.

Rick tried to calm my panicky feelings by yelling to me across the chasm that separated us: "Try to relax!" He reminded me that I had lived 20 years in the searing summers of Arizona and therefore should be used to the heat. Fat chance of that I mumbled sarcastically to no one in particular.

I felt like I had been knocked down by a tsunami of energy. I felt altered beyond words. Waves of energy and information seemed to be pulsing through my body. My heart was racing. We were all cinching up our battery pack leather belts as we lost water weight.

I tried breathing in the hot air slowly to calm my heartbeat down. There was nowhere to go and nowhere to run. I had to pick my way slowly and carefully in the darkness to return to the entrance of this cave.

After finishing up several roles of film and burning up a few batteries for the camera flash, we carefully made our way out of there and back into the tunnels of the main mine.

The mine's tunnels are cooled to a nice temperature of 102

degrees from a couple of airshafts drilled from the top of Naica Mountain. (In 2009, the mine built a new shaft named the Robin's Hole that was over 2,000 feet long.) It felt like air-conditioning. We drank great amounts of Gatorade and other liquids. They were cool to the touch in their aluminum cans at 102 degrees after being in the caves that felt like they were 130 degrees. We were completely drenched in sweat from the super sauna like environment.

I was exhausted when we finished our exploration of the smaller cave but we still had to prepare ourselves to enter the Cave of Crystal Giants to photograph the approximately thirty-six tall selenite crystals.

The latest research shows that the crystals took over a million years to grow from a grain of salt to these immense sizes. Also what was researched about these mysterious selenite crystals in 2013 showed that these crystals held the slowest growth record in nature. The size of a human hair in growth in a century of time!

Waiting for the others, I was sitting down on the ground outside of the cave with my head between my knees. It was then that I started to be concerned that I wouldn't be able to explore the larger cave and see the really gigantic crystals. My head was pounding with a headache and no matter how much liquid I took in I was dehydrated. I prayed for strength and courage. I felt queasy.

I was also scared. There wasn't someone guiding or informing me where to go and how to take care of myself. I had no idea what to expect and didn't feel prepared physically at all for this kind of exploration.

We later learned how dangerous it was to spend over thirty minutes of time in the caves. Fainting from heat exhaustion was a real probability. Brain cells start dying at 107 degrees. And there was no rescue team waiting for an emergency for us outside the caves or in the tunnels of the mine.

It was years later when I learned that real physical harm happens to the body because of these extreme heat and humidity factors. Neurological damage was a major threat, as the body cannot cool down with exposure to 100% humidity. If the body heats to over 104 degrees, there can be the chance of going unconscious. This fact never made itself clear to me when we first explored the caves. Nor was I informed of the physical dangers other than the obvious ones that pertained to climbing.

We all were in real danger yet we overlooked our risks. Foolish on one hand and yet, an opportunity like this will be forever etched in my mind.

Alejandri, the Chief of Security, dressed in coveralls and long underwear underneath, put us all in the truck and drove our exhausted group of explorers closer towards the mining shaft that brings cool air from the surface down into the Naica mine.

We stepped out of the truck and sat down anywhere we could to rest and restore cooler temperatures to the body. This was as close to an acclimation chamber we would ever have and years before more sophisticated technology came to the caves by the La Venta Group in 2006 or the NatGeoTV and Naica Project teams of scientists that rigged a cooling tent called the Ice Cube and created specialized cooling suits and respirators for the scientists.

Laying prostrate, attempting to cool the body down, I almost fell asleep as I looked up at the airshaft bringing some daylight in 900 feet above. I could see the fan circulating inside the shaft that brought the cooler air down to us. Alejandri seemed to be enjoying himself, feeling comfortable at 102 degrees and telling us stories about the Naica mine.

As he sat down next to us and leaned against the mineral encrusted wall of the tunnel, he began like this: "Mexicans have mined here for over two hundred years. What a history this place has! The Cave of Swords was discovered inside the mine in 1910. At that time, at only a level of 400 feet, they were the largest find of selenite crystals. They were only 3 feet long and were bladed. Geologists from all over the world came to Naica to study these crystals because they had been in and out of water as the water table had risen and subsided many times," he said.

"Lead, silver and zinc are the minerals mined for profit but there are a myriad of other minerals as well. Calcite, copper, pyrite, malachite, azurite and some celestite crystals are found here as well." He looked to see if any of us were listening. It was as though he was telling us a bedtime story. I was feeling my eyelids become heavier while the airshaft fan blades continued to whirl above us.

He continued: "In April of 2000, two Mexican miners, the Delgado brothers who are employees at the mine, discovered the crystals while searching in the bedrock for a new main vein of silver. And the rest you could say is now history." He smiled as if to say this was such an easy effort by the miners that of course it wasn't.

It was all very interesting to me as I had never been inside a working mine before, however, I was overwhelmed by my own exhaustion to pay rapt attention to our Chief of Security who seemed to be enjoying the relaxation in the heat. He wore his coveralls and red, long john underwear beneath so that he could manage both being topside where it was about 40 degrees and the hotter temperatures below.

As I marveled at his ability to withstand these grueling temperatures, the worst was yet to come.

CHAPTER 12

The Crystal Cave of Giants

In about an hour, after we recovered from exploring the first cave and the sweat had mostly dried on our skin, we drove about five more minutes down the tunnel to a fork in the passageway. There was a completely sealed concrete wall with a rusted iron door and a heavy lock on one of the openings.

The Cave of the Giants and the Cave of Dreams (later renamed The Eye of the Queen then Queens Eye) by National Geographic and the La Venta teams in 2008) is connected through an adjoining wall.

When the news hit around the globe that gigantic crystals were found in the mine, a few of the miners were attempting to remove the crystals and sell them on the black market in Mexico. However, the weight and enormity of the crystals makes that almost impossible and dangerous to attempt.

One man had already died in the Cave of Giants by using archaic means to move the crystals out of there. He didn't have enough time and lost consciousness. When they finally found him, he was cooked to death. I have some of the earliest images of the caves where his rope was still hanging from the top of one giant pylon. Management decided to increase security and in those early days, the use of an 1800's jail door seemed just about as good as any to keep intruders and thieves away.

When Alejandri keyed the lock and opened the heavy iron jail door, a blast of black, humid, hot air like a super hot sauna hit our faces. It almost knocked me over, not from force but from the feeling of initial danger of intense heat. I was still weak from heat exhaustion from our earlier exploration. At this point, we had not even approached what was then named the Eye of the Queen, otherwise known as the entrance into the cave. In this location, the mineworkers punched a hole as the first opening into the bedrock. On the other side of this hole, a wall of selenite crystals shimmered in the flash of the camera's flashbulb.

Walking another twenty feet and up three concrete steps, we stepped into an opening that resembled a huge eye set in selenite crystals.

The heat was completely oppressive. It was by our best guess 128 degrees inside and 100 percent humidity (in 2001). I grabbed a handkerchief soaked in cool water and put it over my face.

Alejandri guided us in. What I saw was unbelievable! It was truly an alien world never seen on Planet Earth. Stepping inside a giant geode, there were massive structures, 30 to 40 feet tall of solid selenite crystals. They crisscrossed from every direction with one massive one on the floor. (In 2013, we now know these giants are over 1 million years old.) It felt as if I had entered a dream. I was overcome with being completely altered from their energy, feeling my overtaxed body and the dehydration setting in.

I knelt down and laid my hand on one of the very warm solid planks and/or pylons. It felt as solid as steel and as alive as living tree trunks. Nothing in my reality had prepared me for such a place. I was astounded by it.

At this early time of discovery, this place was a complete mystery, as only a few people knew very little about it. What really occurred in this place to form these massive crystals? We had none of these answers and no one would know much more for at least another 6 more years.

In 2008, National Geographic Television presented and aired their first documentary about the Caves from explorations by a team of scientists. Then "NatGeoTV" presented the second exploration in 2009 and 2010 to the world with the second documentary airing in the fall of 2010.

In 2001, we attempted to explore as much ground as possible but the cave was deeply recessed and dark. Not only that, we had no protective gear or clothing to help us stay in the caves longer. It was only years later, when the engineers and miners were able to backlight some of the crystals for newer exploration teams and scientists to spend more time in the caves. They would only be able to do this if they wore the specialized suits and respirators that were being developed years later after we, the first explorers, went into the caves. No one had thought about what it would take to extend our stay in those caverns back in 2001. Our research team certainly underestimated the risks we took with our lives for the goal of exploring the entirety of the caverns.

We had to hurry! The heat made it impossible to stay in there more than five minutes at a time. The challenge was to navigate almost in complete darkness except for the light coming from the helmets we wore. We had no idea then that the cavern was later estimated to be the dimensions of a football field in width and the height of a four-story building.

Having to leave so quickly to acclimate our body temperature did not give us much time to explore deeper into the recesses of the cavern. This posed quite a challenge. How was one to go further into the cave from the last time in the same amount of minutes and yet cover new distance?

I had to return three more times into the Cave of Crystal Giants because a human could not stay conscious in there for more than five or six minutes at a given time without dropping from the extreme heat.

We wanted to explore as much of the crystal room as we could against a clock that was ticking dangerously towards compromising our lives. Again, the challenge was overcoming tasking environmental conditions and crawling over sharp brittle crystal in almost total darkness.

Imagine someone saying to you, explore a football field in its entirety within five minutes or you will not come out alive? No wonder specialized suits had to be invented as late as 2007-2008. These "low-tec" suits consisted on a jumpsuit, mask and enough room to put a vest on underneath that held a dozen cylinders of frozen ice to cool the back and front of the chest. This protected the organs from overheating the body. The many scientific teams that came to this place had to take their time to find out the answers to their questions of alien life from bacteria, age of the crystals and determine if this was the only pocket or would there be more caverns beneath the Naica Mountain Range.

In order to reach an area where I would later describe a gathering of lodge-poles of crystals, we would have to scramble across flanged sharks like teeth crystals that are now known to be the largest formation of desert flower selenite crystals in the world. They were extraordinary and quite beautiful in their almost transparent clarity and gigantic! What we were looking at resembled a prehistoric environment where everything was huge, unknown and felt life threatening. This was something so completely alien for us, for modern-day humans, to see and experience. The flowers were in sizes of 4 feet to 8 feet tall from the floor of the cavern. We had to walk slowly and choose our way carefully as they were sharp and pointy and susceptible to breakage.

Crawling up and over these points proved to be tedious and exhausting and chewed up precious time on the clock to explore further into the reaches of the giant crystal cavern.

Once over these giant flowers, we walked across a crystal plank (this is the largest one that had fallen in the caves) where we could stand or kneel under several giant pylons overhead. Rick witnessed me in complete awe as I looked above my head and reached to touch this one crystal that seemed completely perfect in proportions and clarity. And that is where my photo was shot for the April 2002 issue for the Smithsonian Magazine. This article indeed implicated to the world that history was being made by the discovery and sheer size of these crystals as to their being the largest in the world. Before this discovery, there had not been substantive proof to make such a bold statement and so there was a waiting period of two years before such claims were being made globally. The article is provided below: (These earliest comments are not correct in their total description)

Crystal Moonbeams
"A pair of Mexican miners stumbles upon a room filled with what could be the world's largest crystals"

By John F. Ross
SMITHSONIAN MAGAZINE APRIL 2002

Deep below the surface of an isolated mountain range in Mexico sit two rooms of splendor: translucent crystals the length and girth of mature pine trees lie pitched atop one another, as though moonbeams suddenly took on weight and substance.

In April 2000, brothers Eloy and Javier Delgado found what experts believe are the world's largest crystals while blasting a new tunnel 1,000 feet down in the silver and lead Naica Mine of southern Chihuahua. Forty-year-old Eloy climbed through a small opening into a 30- by 60-foot cavern choked with immense crystals. "It was beautiful, like light reflecting off a broken mirror," he says. A month later, another team of Naica miners found an even larger cavern adjacent to the first one.

Officials of the Peñoles Company, which owns the mine, kept the discoveries secret out of concern about vandalism. Not many people, however, would venture inside casually: the temperature hovers at 150 degrees, with 100 percent humidity.

"Stepping into the large cavern is like entering a blast furnace," says explorer Richard Fisher of Tucson, Arizona, whose photographs appear on these pages. "In seconds, your clothes become saturated with sweat." He recalls that his emotions raced from awe to panic.

Fisher says a person can stay inside the cave for only six to ten minutes before becoming disoriented. After taking only a few photographs, "I really had to concentrate intensely on getting back out the door, which was only 30 to 40 feet away." After a brief rest, he returned for another couple of minutes. "They practically had to carry me out after that," Fisher says.

Geologists conjecture that a chamber of magma, or superheated molten rock, lying two to three miles underneath the mountain, forced mineral-rich fluids upward through a fault into openings in the limestone bedrock near the surface. Over time, this hydrothermal liquid deposited metal such as gold, silver, lead and zinc in the limestone bedrock. These metals have been mined here since prospectors discovered the deposits in 1794 in a small range of hills south of Chihuahua City.

But in a few caves the conditions were ideal for formation of a different kind of treasure. Groundwater in these caves, rich with sulfur from the adjacent metal deposits, began dissolving the limestone walls, releasing large quantities of calcium.

This calcium, in turn, combined with the sulfur to form crystals on a scale never before seen by humans. "You can hold most of the crystals on earth in the palm of your hand," says Jeffrey Post, a curator of minerals at the Smithsonian Institution. "To see crystals that are so huge and perfect is truly mind-expanding."

In addition to 4-foot-in-diameter columns 50 feet in length, the cavern contains row upon row of shark-tooth-shaped formations up to 3 feet high, which are set at odd angles throughout. For its pale translucence, this crystal form of the mineral gypsum is known as selenite, named after Selene, the Greek goddess of the moon. "Under perfect conditions," says Roberto Villasuso, exploration superintendent at the Naica Mine, "these crystals probably would have taken between 30 to 100 years to grow."

Until April 2000, mining officials had restricted exploration on one side of the fault out of concern that any new tunneling might lead to flooding of the rest of the mine. Only after pumping out the mine did the level of water drop sufficiently for exploration. "Everyone who knows the area," says Fisher, "is on pins and needles, because caverns with even more fantastic crystal formations could be found any day."

Previously, the world's largest examples of selenite crystals came from a nearby cavern discovered in 1910 within the same Naica cave complex. Several examples from the Cave of Swords are exhibited at the Janet Annenberg Hooker Hall of Geology, Gems, and Minerals at the Smithsonian's National Museum of Natural History."

Photo used for Smithsonian Magazine 4-2002

It was only much later; we were to clearly realize that these were the largest crystal caves to be found on planet earth when the other teams of Scientists came to Naica starting in 2006.

Further into the cave, with only a couple minutes left, we encountered huge crystal trunks the size of giant Sequoia trees. One column was six feet wide of pure crystal. That is when I found the dead man's rope that was used in his attempt to steal one of the smaller crystals hanging from the roof of the cavern.

I am sure this tactic was used when the Cave of Swords was discovered in 1910 because I have seen several of these blades for sale over the years at some of the gem and mineral shows around the U.S. The blades were only three feet long. However, these crystals proved to be resistant to breakage and the thieves ran out of time before they could run away with a crystal outweighing them.

Looking at the rope in this mind-bending environment made me feel I was watching a sci-fi movie. Viewing the slides/photos later on, my biggest surprise was to see we were dwarfed by these huge sentient beings that seemed to pulse unknown and mysterious messages of their origins.

Also, it seemed to me that there was an indistinct pale blue light emanating from and through them and it wasn't coming from any reflection of our clothing.

When we could no longer take the heat, we stumbled out of there, blinded by the sweat in our eyes, completely drenched and exhausted. When we felt somewhat recovered, another courageous trip to return to the caves was inevitable. Our curiosity was undeniable and it pushed us beyond our boundaries for safety.

We all had easily lost five pounds of water weight by noticing that we had to continually cinch our battery pack belts tighter and tighter around our waists.

Hurrying out of the cave as fast as I could after each visit brought frustration to our photographer. We shot in rapid fire many photographs in a short duration of time. I stumbled and crawled and explored as much as was possible in this cavern. I would call this crawling around in almost pure darkness, my "feeling in the dark" moments as I searched for crystal treasures. Then the feeling of dizziness would hit and my body would signal me to make a move towards the opening and force me to leave immediately.

It was almost impossible to stay a second longer. Being completely drenched by sweat and humidity, my skin now was as red as a lobster.

Alejandri just kept his so-called cool through the entire exploration, as he was our only security and rescue team in the caves with us. However, I will never forget when he took me by the hand and coaxed me in a little further one last time; and said in Spanish: "You must look and remember. Remember! Remember! Yes?" I could barely acknowledge his words from sheer exhaustion. Mercifully, when he would not press any further; I gave him an answer of a complete and solemn promise that I would never forget. (Now, knowing these giants are again submerged under water as of 2015, I am grateful I heeded his admonishment.)

All of us had wanted a few pieces of the sparkling crystals of selenite to take with us as a remembrance of our exploration. To our surprise, even with the multitude of crystals all around us, we were told no, there would be no exceptions allowed in taking crystal out of the cave.

Imagine the disappointment and my determination. This just couldn't be the only answer and so I asked for a most benevolent outcome to this situation.

Grateful at that point to be leaving the cave, I stumbled out of the heat into the cooler temperatures of the tunnels. As we waited for each one of our team members to complete their discovery, we took one last photo of the exhausted explorers.

At this point, my head was in excruciating pain. I was having a migraine that was blurring my vision. I knew that I was on the brink of dehydration and did my best to drink the hot Gatorade that the mine provided us.

I was mostly silent from my horrendous headache that was beginning to overcome me as we made our way into the security truck and drove slowly up to the top of the mine where we went into locker rooms to change back into our street clothing.

Speechless and with deep gratitude, I realized no one in our group had experienced any injuries. It took us a while to recover before we went to our vehicles and drove back to the City of Chihuahua.

Our bags were searched thoroughly before we left the premises. It was at this point, Alejandri gave us a few pieces of selenite crystal that were small enough to place in our hands as gifts directly from the caves. I was overcome with joy and thanked him with all sincerity as I had asked for a benevolent outcome for this request.

We drove out on a bumpy dusty road back through the town of Naica. My head was pounding from the dehydration causing me to have a massive headache now at the point of just wanting to lie down and close my eyes.

As incredible as it seemed, I actually fell asleep with my arm on the door holding my head in my right hand. I was praying for a complete recovery and within minutes, after falling deeply asleep, my intense migraine was gone. It was a miraculous recovery for me. I was humbled by the swift miracle of healing.

We entered the town of Delicias and had cold beer and Mexican food for dinner before returning to the Palace of the Sun Hotel in Chihuahua.

This indeed was one of the most mind bending, exciting and amazing experiences of my life, and it was one of the most challenging. It is hard to believe now that all of us risked our lives to some extent to explore these caverns for the State of Chihuahua government and tourism and the mining corporation with our reports.

There is mystery and magic of things discovered and yet unexplainable on this earth by humans. There are no other giant crystals from minerals that have been found on earth that we know of that can be viewed or explored but we do know they exist. And as far as we know in 2016, nothing comes close to being found in or on the planet that compares to the enormity of the selenite crystal giants!

On a more cosmic note, I question at times how I was one of the first to explore the crystal caves? And where will this adventure lead me next? Singapore and Malaysia in 2016 seem to be on my radar.

At the time I went into the caves, I did not know about being one of the first explorers. I have since learned that the Penoles Mining Group had allowed no women to enter the mining operations (up until January of 2001) because of a superstition. For Sonia Morales, a representative of the Chihuahua State government and I to be the first women to enter these caves seems astonishing and an honor to say the least.

This was a private government expedition for the purposes of exploration. To my knowledge, the public has not been allowed for to enter and marvel at these extraordinary crystals, as far we know. The intense heat with 100% humidity is a deadly combination and this mine is named the most hostile environment in the world to work in.

In 2015, we now are informed the mine has shut down the turbine engines that pump the aquifer water table to a level of 2,200 feet below the surface. The decision had to do with how much ore can be extracted from the mine to make a profit. It is at a huge expense to run the pump's engines twenty-four hours every day to pump over a million gallons per hour out of the mine. Sadly, the day has come where the mine's vast tunnels are flooded once again and the pockets of crystal covering up the silent giants are now forever out of reach.

Today, in 2016, and in years past, I continue to present the earliest and rare photos, give workshops, lectures and interviews about the mysterious giant selenite crystals. It has been an honor and a privilege to share this amazing discovery (from my perspective and personal journey) with excited audiences around the globe. Many are drawn and are interested in the deeper knowledge of how these particular crystals are connected with us, as humans on this planet.

Mother earth created these crystal earth-keepers. I don't believe anything just happens without a purpose or function, even the growth of crystals. With further research in and around Naica, there are other paranormal events linking the mystery of giant crystals that seem to be unexplainable, such as giant human bones found in the mountains.

CHAPTER 13

Summary of the Into the Lost Crystals Documentary
October 2010

With regard to the latest research of the Naica giant selenite crystals, produced and aired by National Geographic TV, (entitled Into the Lost Crystals) in 2010, there has been the discovery of more crystal caves that indicate there are connecting passages.

Since 2001, when I visited the mine and began my own research into understanding the incredible phenomena of giant crystals, I believed that this system not only was centered in and around the first pocket of giant crystals but that a sub-channel of gypsum crystals (a crystal river) would lie on a northeast-southwest axis. This sub-channel would surface further north towards the greater White Sands National Monument in southern New Mexico.

In January of 2008, my travels to New Mexico returned me to my childhood memories of living in near-by El Paso, Texas. Student field trips took us to the Carlsbad Caverns and the very first experience I saw of living crystal was deep into those caverns.

As we drove eastward from El Paso that morning, towards Carlsbad, I had asked my friend Perry to pull over on the side of the highway and look southward over a vast portion of the desert called the Delaware Basin. It exposes part of a fossilized reef called the Guadalupe Mountains. The Delaware Basin is part of the larger Permian Basin that covers over 10,000 square miles of Texas and southern New Mexico.

This period of deposition left a thickness of 1600 to 2200 feet (490 to 670 m) of limestone inter-bedded with dark-colored shale. Limestone is a sedimentary rock composed largely of the minerals; calcite and/or aragonite, which are different crystal forms of calcium carbonate. Like most other sedimentary rocks, limestone is composed of grains; however, most grains in limestone are skeletal fragments of marine organisms such as coral or foraminifera. Other carbonate grains comprising limestone are ooids, peloids, intraclasts, and extraclasts.

Some limestone does not consist of grains at all and is formed completely by the chemical precipitation of calcite or aragonite. i.e. travertine. The solubility of limestone in water and weak acid solutions leads to karst (cavern-like) landscapes. Regions overlying limestone bedrock tend to have fewer visible groundwater sources (ponds and streams), as surface water easily drains downward through joints in the limestone. While it is draining, water and organic acid from the soil slowly (over thousands or millions of years) enlarges these cracks dissolving the calcium carbonate and carrying it away in solution. Most cave systems are through limestone bedrock.

It was in May of 2008, while I was on a crystal dig in the western part of Arkansas, that I found out my theories were correct when I met the new owners of an old military controlled deposit of gypsum (Selenite crystals made from calcium carbonates) near Carlsbad, New Mexico, not 2 miles from where I had stopped our vehicle 5 months earlier.

We made an immediate connection upon our meeting and an invitation to visit the deposit was forthcoming. It was in December of 2009, that I had the opportunity to spend a couple of days at the Selenite deposit. My time there, especially spending the night in the site trailer set me up for some extraordinary dreams (Angels or angles of light appearing to me and the context for healing my body through the laying of their hands or energy connection to my feet).

Upon closer inspection of the crystals, the quality of the material was equal to the clarity and transparency of the crystals featured in the two documentaries (October, 2008 and 2010) by National Geographic on the largest Selenite crystals ever found on the planet. The National Geographic documentary is focusing upon its final expedition into the caves to search for more crystals and rare living microorganisms not found anywhere else on earth.

Since the conditions are life threatening, exploring in the caves at 118 degrees (now 113 degrees) of heat with 90 percent humidity can kill a human within 30 minutes. The extremely hot temperature of the mine has been slowly dropping since the new ventilation shaft drilled 2,000 feet down into the mountain in 2009 and the opening of the caves allowing the steam and heat to dissipate. Many people will risk their lives to understand the creation of these naturally formed crystals and their impact on earth and humans.

A visual marvel unlike any other, a crystal cathedral defies our understanding of what earth is capable of creating. An alien world right here on earth, the caves don crystals covering the length of a football field and the height of a two story building.

Beyond the Plexiglas door, near the tunnel access leading to the caves, there is a forest of crystals and passageways that have not been fully explored by the Scientists in 2010.

Penny Boston, Astro-Biologist from New Mexico Tech has returned from the 2008 documentary. No one else has returned from the original team. She has returned to find ancient viruses trapped inside water crystals. Looking for other life forms not previously discovered before may help in our understanding of how life may have formed and/or created on other planets.

The teams have assembled a see through cooling structure called the ICE CUBE that was designed to help the crew recover quickly from over-exertion to the heat and humidity.

In 2008, the explorers were proving there were other pockets of crystalline structures that exist. Any time over one half hour of exploration can be deadly.

On top of the Naica Mountains, to the west, another team is assembling a belay to repel down an air ventilation shaft drilled by the mine owners in 2009 to help the miners stay cool in the extreme heat that occurs in the mine. This shaft goes to down 2,000 feet and is called the Robin's hole.

Mark Beverly, the leading explorer repelled down 500 hundred feet to where there was a pocket broken into the shaft through the bedrock. There lies a pocket of crystals in an unexplored cave they named the 'Ice Palace', but nothing the size of the Cave of Crystal Giants.

A volcanic intrusion of magma created the extreme heat conditions that heats up an underground aquifer. Thus, the experience of extreme heat and humidity are what the miners must endure during exploration.

There are one million gallons of water pumped out of the mine per hour. In 2008, the officials of the mine said that the turbines are to shut down in the future months thus ensuring the crystals will be underwater once again making sure they will be forever out of our reach. (This has now happened since 2015)

Since further exploration is impossible beyond thirty minutes, air-conditioned suits were designed to extend the safety and wellbeing of the crews who would look further and longer into the caves.

Photos were also a big issue as cameras could fog up in a moment and then malfunction. They were able to resolve the problem by placing cameras for filming and photos in plastic bags and exposing the equipment to the extreme conditions for 2 hours. Thus, the cameras were acclimated to the temperature and humidity.

There was a newer crevasse discovered in 2006 with attempted exploration in 2008 that leads to another cave of crystals. However, even though there is cooler air moving through a passageway, there is no easy access.

The Selenite crystals were formed from gypsum, an evaporite made from salts, more specifically, hydrated calcium sulfate that covered the walls of the caves underwater. With the perfect conditions, the crystal's growth occurred under water and became gigantic.

Penny Boston and her co-worker found enhydro crystals (water bubbles) within the Selenite crystals inside the QUEENS EYE cave. Their mission was to find new microorganisms, specifically, bacteria. They reanimated one sample of bacteria chewing up the bedrock using mineral compounds as an energy source. This specific bacterium has not been found before on Earth.

Penny became distraught by the heat. Overwhelmed with her emotions, she started to cry and her comments regarding her exploratory work in the crystal caves is that these crystals are Earth's gift to us and we may not see them much longer. Her forecast was correct.

In the 1910, the very first cave of Selenite crystals was discovered at the Penoles mine in Naica. Geologists from all over the world would come to Mexico to understand the nature of the creation of them.

This was the Cave of Swords at was found at -200 feet below the earth's surface. The oscillation of water that rose and fell many times was the reason why these particular crystal specimens did not grow to such gigantic proportions but became somewhat stunted. At -900 feet, the giant crystals that grew in perfect conditions of water did not oscillate. This is a clue to more caves.

On the surface, Mark Beverly explores the Robin's Hole cave at 500 feet named the Ice Palace cave and said the temperature in the air shaft was so hot it felt radioactive.

A new cave, the Cave of Sails was discovered in 2009 after drainage of water over 20 years ago and its underground location revealed new structures of Selenite in long needles, broccoflowers (broccoli and cauliflower) and nested needles. Upon further examination, the nested needles are from aragonite. Aragonite is a carbonate mineral, one of the two common, naturally occurring crystal forms of calcium carbonate, $CaCO_3$ (the other form is the mineral calcite.) Crystals are formed by biological and physical processes and include precipitation from marine and freshwater environments.

Although there is no mention of the massive aquifer underground, its implications are major when it comes to understanding why the crystals are formed here in the dry high desert elevations of Mexico. There has been no outlet for the massive amount of water that found its way to the Gulf of Mexico via the Rio Grande during the time when Oceans were formed in this region.

In 2001, being one of earliest explorers, we spent a lot of time in the smaller cave then known as the Cave of Dreams. This cave has been renamed Queen's Eye. There was some question this cave is connected with the Cave of Giant Crystals in 2001. There are crystal structures here that are not found anywhere inside the larger cave. Namely, the gigantic desert rose or flower that selenite often forms in. These are exceptionally clear crystal instead of full of sedimentary material normally seen in mineral collections.

In addition, there is a mass amount of water remaining at the lowest levels of the mine since all of the water is difficult to pump out with replaced water coming back in from the surges of flow from the underground aquifer.

Viruses are predators and eat bacteria. Test samples of the water in the Queen's Eye cave reveals 10 million viruses within 1 millimeter. There is a question concerning the connection of these viruses to the same life forms that are at the bottom of the ocean found in volcanic steam vents. The question perplexing the scientists now is how did deep oceanic microorganisms find their way to the waters that formed the crystal caves in Naica?

There were three new microorganisms discovered at the lowest region of the mine that have not been found on earth before. Could this possibly help us understand the alien bacteria that fell from meteors and comets that impacted earth?

La Venta, the Italian team that is responsible for the exploration of the caves stated they would soon be flooded. (As stated, the caves were flooded in 2015.) The strange formation of crystals will become a heritage to our understanding of life in our solar system, not just planet earth.

During my travels to the Southwestern part of the United States for the past 15 years, I have researched an unusual amount of "high strangeness" that is connected to the Chihuahua desert in and around the location of the Penoles Mine in Naica where the crystals were found. Could there be alien life connected to these crystals?

New Speleothem discovered in Caves

Stay tuned for information that I will be writing in my second book and more posted on my website at www.thecrystalgiants.com and Facebook at: https://www.facebook.com/giantcrystalsofmexico and https://www.facebook.com/naicacaves

CHAPTER 14

Reflections

Approximately one year after being in the caverns in Naica, a woman sent me a poem from Silver City, New Mexico. She had heard of my adventure and wanted to share with me what the Giant Crystals were speaking to her. They had a message for me. I was very touched by the depth of meaning in this poem.

After I received it in an email from her, plans were made and I drove to Silver City to meet and spend some time with her. We spoke of many things as we hiked and picked up some unusual rocks and even Indian arrowheads in the area and recognized that she was a long-time crystal sister.

I would like to share this poem with you, the reader:

The Giant Crystals Speak
By Marilyn Twintrees
Silver City, New Mexico

Today is a breath in the void

A most momentous now

Greetings to all who can hear, all who can breathe and smell the ancient waters we are born out of

We emerge from the ancient aquifer of water fire, the spark of life itself encoded in every droplet of earthly waters

If you know of us, then you know of the you that you are becoming

The evolution that thrives in your spirit so perfectly that it lands upon your body

Crystallizing into the most exquisite union of dream and peace

Breathe this in fully

You have brought it to yourself with sweet intent and free power

You have discovered us

In doing so, we tell you that you have discovered for yourselves

The core of your infinite connection to the earth and to life through love

We are the core of the earth

We are the core of Love meeting form and recognizing itself as pure divinity

In your mind, you have thought yourself separate

The need for this has dropped away

In our presence you witness the magnificence and it is we, and it is you

It is we who are joined and intermingled in this season of growth

To see us is to know your dreams fulfilled

Not just what you think you want
But that, what your soul has placed into your outstretched heart

Because it is truth

Because it is meant to be

In freedom

When you gaze upon us your cells understand this is the time

That you knew of at your own core

We have synchronized our beings

Now your DNA knows to activate its full genius

Now you remember the complete wonder of you

A breath

At a time

We meet you in that breath

Now

ABOUT THE AUTHOR

Leela Hutchison is a Graduate Gemologist from Gemological Institute of America in Carlsbad, California. As an explorer, teacher, presenter and now author, on crystals, gems and minerals, she specializes in educating listeners on the remarkable qualities of Selenite, considered by many to be one of the major power generating energies of the emerging new global consciousness.

She began showing rare images of the earliest exploration of the Giant Selenite Crystals found in Mexico starting in May, of 2001 to audiences on the island of Kauai. From the overwhelming interest from that audience, invitations were extended to teach around the globe.

Leela is also a healing arts practitioner since 1996 with more than 5,000 hours of hands-on-healing. She specializes in crystal energy amplification with the use of rough and faceted gemstones, quartz and Selenite crystals. Her classes provide education to utilize these same healing modalities through the layout of grids of crystalline energy patterns on the body and energy gridlines on the earth.

She was born and grew up in El Paso, Texas where her love for exploration and collecting rocks began. A fascination with the geology of the American Southwest led to what she laughingly refers to as a near-obsession for hiking into canyons, mountain ranges and caves of the region with an endless curiosity for exploration and rock hunting.

For years, Leela has traveled areas of the world to power centers and sacred sites including Peru, England, Bermuda, Iona, Scotland, Greece, Mexico, Arkansas, Ireland, Canada and France. Last year while visiting Portugal and Spain, she experienced more evidence of telluric energies and gridlines running through and under the Pyrenees Mountains that border with France. She also visited the pyramids of the Yucatan and Oaxaca, Mexico, in pursuit of scientific, spiritual and metaphysical understanding of energy in stones, vortices, key Ley lines and grid lines in the earth.

In 1997, she experienced her first awakening to the greater unified field of consciousness, and the realization of the great love and honoring that mother Earth has for all of her children.

In 2001 Leela became the first American woman to enter the astonishing giant Selenite crystal caves near the village of Naica, Chihuahua in the Sierra Madre Mountains of Mexico. These caves contain what are now known as the largest crystals on Earth ranging in size to almost 15 meters or 36 feet in height, weighing as much as 60 tons and estimated to be 1,000,000 (one million) years old.

The conditions of the cave are one of the most hostile environments on earth: extreme temperatures of 132 Fahrenheit in 2000 and now 113 degrees (in 2010 and slowly dropping) with100% humidity then and now 90%.

Leela feels intuitively there are still more of the colossal Selenite crystals yet to be discovered deep underground. For the past fifteen years, she has been gathering research suggesting their energy fields affect the collective consciousness of humanity. Her research includes on-going explorations linking these giant Selenite crystals to other Selenite crystal deposits found in New Mexico and unusual landmarks as far as Jalisco, Mexico, southern Colorado, Florida and soon Baja California.

Linking grid lines from power spots or sacred sites on the earth to Naica is also part of her teaching to students interested in crystal grid technologies.

In mid-December, 2009, while boarding a plane in Los Angeles to visit west Texas and the Permian basin, she had a conversation with New Mexico Tech associate professor of Cave and Karst Science, and Director of Cave and Karst Studies, Department of Earth and Environmental Science, Astro-Biologist: Penelope Boston. Penny had recently returned from Naica on another expedition with the Naica Project team.

They had just finished filming Part 2 of the Naica Project for National Geographic television. "NatGeoTV", filmed part 1 of the Naica Project for their documentary in November of 2008. Part Two was aired in October of 2010.

Their team had found another pocket of the Selenite crystals in the Naica Sierra Madre mountain range. Although these crystals are not in the same giant formation as the first cave found it is an indication of a mass amount of material found underground in this area.

They also discovered a very unusual speleothem never seen before. (Please watch "Into the Lost Crystals" documentary by National Geographic Television for further in-depth details.)

Leela's latest research is now revealing how these crystals amplify the crystalline field found in the electromagnetic ley lines in and around the earth. These selenite transmitters have connected their messages into the trigger dates of activating the 144 vertices of the crystalline grid. (The double penta-dodecahedra) The last trigger date of 12-12-12 was the culmination of all of the major planetary crystals coming on line for the completion date of the historic December 21st, 2012, the winter solstice sun and the galactic alignment.

She currently resides in the Valley of the Moon in Sonoma County of Northern California. Writing her first book here has given inspiration to compile all the research gathered for her second book soon to be published in 2017.

For scheduling a phone consultation for a session in gem/crystal energy healing, or birthing and navigating your path for your true purpose and creativity and/or counseling on specific gems and minerals and their appropriate use: she can be reached at 001-415-847-0141 if calling outside of the United States.

If requesting workshop, lectures or presentations, please reach me at leelasgems@yahoo.com.

LEELA HUTCHISON

www.ingramcontent.com/pod-product-compliance
Lightning Source LLC
Chambersburg PA
CBHW040808200526
45159CB00022B/50